Cambridge
International AS and A Level Mathe

Pure Mathematics 2 and 3
Practice Book

Greg Port

HODDER
EDUCATION
AN HACHETTE UK COMPANY

Answers to all of the questions in this title can be found at www.hoddereducation.com/cambridgeextras

Questions from the Cambridge International AS and A Level Mathematics papers are reproduced by permission of Cambridge International Examinations.

> Cambridge International Examinations bears no responsibility for the example answers to questions taken from its past question papers which are contained in this publication.

This text has not been through the Cambridge endorsement process.

Questions from the NZQA Mathematics papers are reproduced by permission of the New Zealand Qualifications Authority.

(1) the New Zealand Qualifications Authority owns the copyright in the examination material reproduced and has consented to the reproduction of that examination material
(2) any component of the examination material in which the copyright belongs to a third party remains the property of that third party, and has been used in the publication with their permission.

Every effort has been made to trace all copyright holders, but if any have been inadvertently overlooked the Publishers will be pleased to make the necessary arrangements at the first opportunity. Although every effort has been made to ensure that website addresses are correct at time of going to press, Hodder Education cannot be held responsible for the content of any website mentioned in this book. It is sometimes possible to find a relocated web page by typing in the address of the home page for a website in the URL window of your browser.

Hachette UK's policy is to use papers that are natural, renewable and recyclable products and made from wood grown in sustainable forests. The logging and manufacturing processes are expected to conform to the environmental regulations of the country of origin.

Orders: please contact Bookpoint Ltd, 130 Milton Park, Abingdon, Oxon OX14 4SB. Telephone: (44) 01235 827720. Fax: (44) 01235 400454. Lines are open 9.00–5.00, Monday to Saturday, with a 24-hour message answering service. Visit our website at www.hoddereducation.com.

© Greg Port 2015

First published in 2015 by
Hodder Education, an Hachette UK Company,
Carmelite House, 50 Victoria Embankment,
London EC4Y 0DZ

Impression number 5 4 3 2
Year   2019 2018 2017 2016

All rights reserved. Apart from any use permitted under UK copyright law, no part of this publication may be reproduced or transmitted in any form or by any means, electronic or mechanical, including photocopying and recording, or held within any information storage and retrieval system, without permission in writing from the publisher or under licence from the Copyright Licensing Agency Limited. Further details of such licences (for reprographic reproduction) may be obtained from the Copyright Licensing Agency Limited, Saffron House, 6–10 Kirby Street, London EC1N 8TS.

Cover photo © Irochka – Fotolia

Illustrations by Datapage (India) Pvt. Ltd
Typeset in 10.5/14pt Minion Pro Regular by Aptara, Inc.
Printed in Great Britain by CPI Group (UK) Ltd, Croydon, CR0 4YY

A catalogue record for this title is available from the British Library

ISBN 978 1444 196320

Contents

P2 Pure Mathematics 2

Chapter 1	Algebra	1
Chapter 2	Logarithms and exponentials	19
Chapter 3	Trigonometry	38
Chapter 4	Differentiation	70
Chapter 5	Integration	111
Chapter 6	Numerical solution of equations	130

P3 Pure Mathematics 3

Chapter 7	Further algebra	144
Chapter 8	Further integration	161
Chapter 9	Differential equations	186
Chapter 10	Vectors	206
Chapter 11	Complex numbers	248
	Past examination questions	289

1 Algebra

Operations with polynomials

EXERCISE 1.1

1 State which of the expressions below are polynomials.

   If they are polynomials, state the order of the polynomial.

Expression	Polynomial Yes / No?	Order
$2x - x^2$		
$\dfrac{x}{2} - \dfrac{2}{x}$		
0		
$x^{23} + 2x^{15} + 1$		
$x^3 - 2x^2 + \sqrt{x}$		
$x^2 + \sqrt{2}$		
$5 + x + \pi x^{45}$		
$1 - 3x$		

2 Find the values of the constants A, B, C and D in the following identities.

  (i) $x^3 - 4 \equiv (x-1)(Ax^2 + Bx + C) + D$

  OCR MEI Further Concepts for Advanced Mathematics FP1 4755/01 June 2007 Q3

  (ii) $2x^3 - 3x^2 + x - 2 \equiv (x+2)(Ax^2 + Bx + C) + D$

  OCR MEI Further Concepts for Advanced Mathematics FP1 4755 June 2006 Q2

3 Simplify these expressions as much as possible.

  (i) $(3x^3 + 4x^2 - 2x - 1) + (x^3 - 2x^2 + 7x + 1)$

  (ii) $(3x^4 - x^3 + 6x - 11) - (2x^4 + x^2 - 4x - 12)$

  (iii) $(x^3 - 5)(3x^3 - 2x + 1)$ (iv) $(x+3)(x^2 - 2)(x^3 - 4x)$

  (v) $(x^2 - 1)^2 - (x^2 - 2)^2$ (vi) $(x+2)(x^3 - 1)^2 - x(x^3 - 1)$

4 Use long division to divide the polynomials, giving the quotient and remainder.

(i) $x+2 \overline{)x^2 + 3x - 1}$

(ii) $x-1 \overline{)x^3 - 4x + 3}$

(iii) $2x+1 \overline{)2x^4 - x^3 + 15x^2 + 12}$

(iv) $3x-2 \overline{)9x^3 + 17x + 8}$

(v) $\dfrac{3x^2 + 5x - 1}{x^2 + 1}$

(vi) $\dfrac{4x^3 - 7x^2 + 8x + 14}{x^2 - 2x + 1}$

5 Find the quotient and remainder for the division $\dfrac{x^4 + x^3 - 3x^2 + 2x - 4}{x^2 - x}$.

6 In the division $\dfrac{x^3 + 5x + a}{x^2 - b}$ the remainder is 2. Find the values of a and b.

Solution of polynomial equations

EXERCISE 1.2

1 (i) Find the remainder when $f(x) = x^3 - 3x^2 - 6x + 8$ is divided by each expression.

    (a) $x + 1$ **(b)** $x - 3$

    (c) $2x + 1$ **(d)** $x - 1$

(ii) Using your answers to part **(i)**, fully factorise f(x).

(iii) Hence solve f(x) = 0.

2 (i) Show that $x + 1$ is a factor of the polynomial $h(x) = 2x^3 - 5x^2 - 4x + 3$.

(ii) Hence factorise h(x) fully.

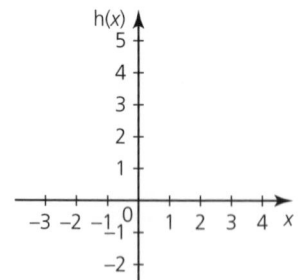

(iii) Solve h(x) = 0.

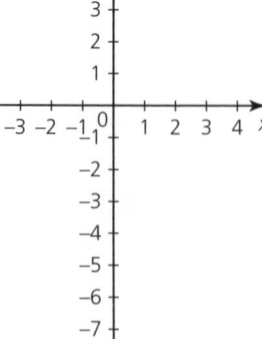

(iv) Sketch the graph of y = h(x).

3 Find the value of the constant p given that $x - 4$ is a factor of $x^3 - 3x^2 + px$.

4 Find the value of the constant m given that $2x - 1$ is a factor of
$f(x) = 2x^4 - 5x^3 + mx^2 - x + 4$.

Hence factorise f(x) completely.

5 When the polynomial $f(x) = x^4 - 6x^3 + 7x^2 + px + q$ is divided by $x + 1$ the remainder is zero.

When f(x) is divided by $x + 2$ the remainder is 72. Find the values of p and q.

6 Find the values of p and q if $x - 2$ and $x + 1$ are factors of $f(x) = x^4 + px^3 + 9x^2 + qx - 12$.

7 When the polynomials $h(x) = x^4 - 8x^3 + kx^2 - 6x - 9$ and $g(x) = kx^3 + 3x^2 + 7x + 13$ are divided by $x + 1$ the remainder is the same. Find the value of k.

8 The polynomial $2x^3 + 6x^2 + ax + b$, where a and b are constants, is denoted by p(x).

It is given that when p(x) is divided by $x + 3$ the remainder is -25 and that when p(x) is divided by $x - 2$ the remainder is 55.

(i) Find the values of a and b.

(ii) When a and b have the values found in part **(i)**, find the quotient and remainder when p(x) is divided by $x^2 + 1$.

9 Let $p(x) = x^3 + ax^2 + 3x$ be a polynomial, where a is a real number.

When p(x) is divided by $x - 2$ the remainder is 26.

Find the remainder when p(x) is divided by $x + 4$.

10 When $x^4 - 2x^3 - 7x^2 + 7x + a$ is divided by $x^2 + 2x - 1$, the quotient is $x^2 + bx + 2$ and the remainder is $cx + 7$. Find the values of the constants a, b and c.

The modulus function

EXERCISE 1.3

1 (i) Sketch the graphs of the following on the grid.

$y = |x + 2|$

$y = |x| + 2$

$y = |2x|$

$y = 2|x|$

$y = -2|x|$

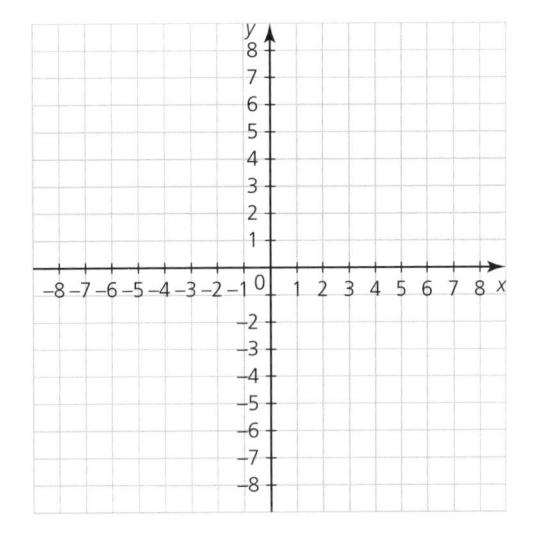

(ii) Describe what effect the constants a, b, c, d and e have in these equations.

Equation	Description
$y = \|x + a\|$	
$y = \|x\| + b$	
$y = \|cx\|$	
$y = d\|x\|$	
$y = -e\|x\|$	

2 Sketch the graphs.

(i) $y = |2x - 1|$

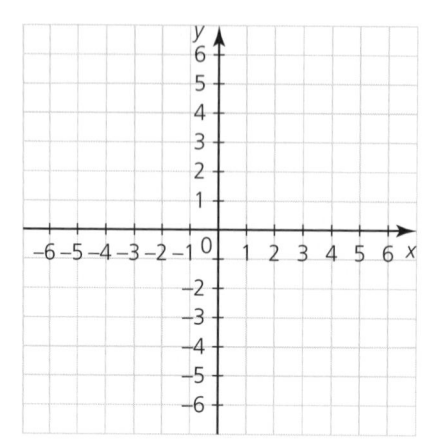

(ii) $y = |3 - x|$

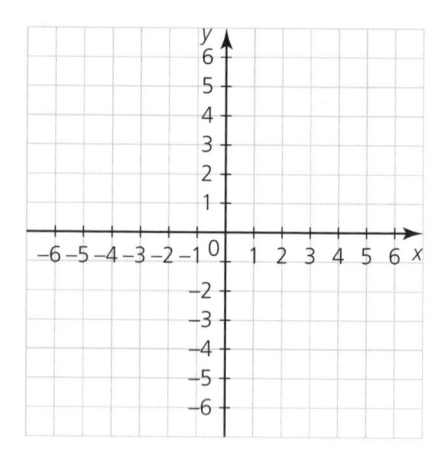

(iii) $y = 2|x + 1|$

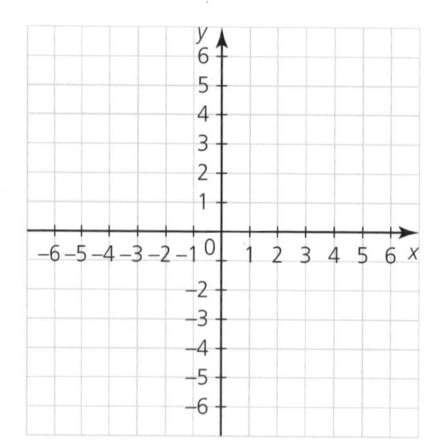

(iv) $y = |x^2 - 1|$

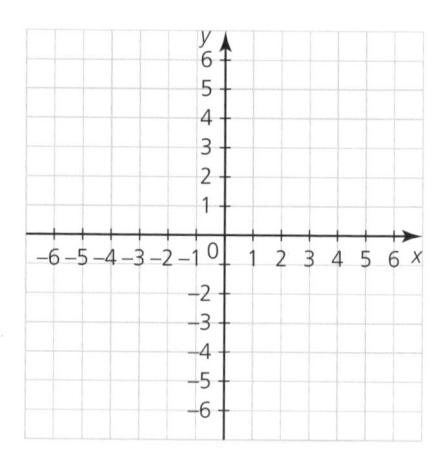

(v) $y = |3 - 2x| + 1$

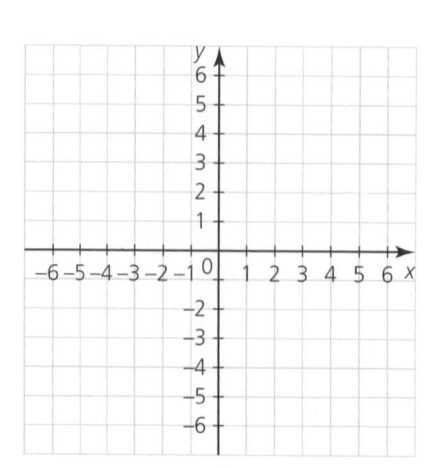

(vi) $y = 2 - |x - 1|$

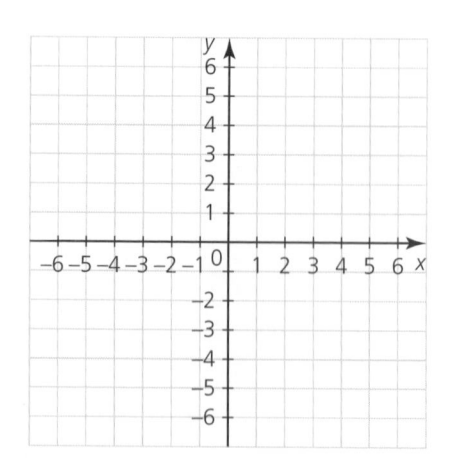

3 The graph of $y = \sin x$ is shown for $-2\pi \leq x \leq 2\pi$.

On the axes sketch the graphs of the equations given.

(i) $y = |\sin x|$

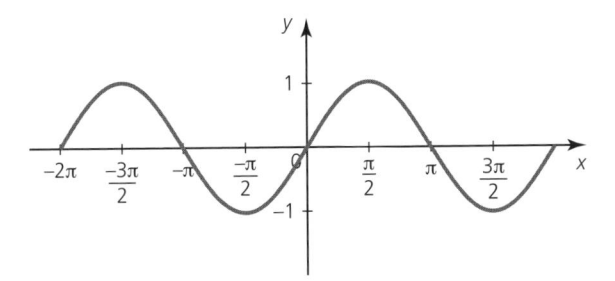

(ii) $y = \sin|x|$

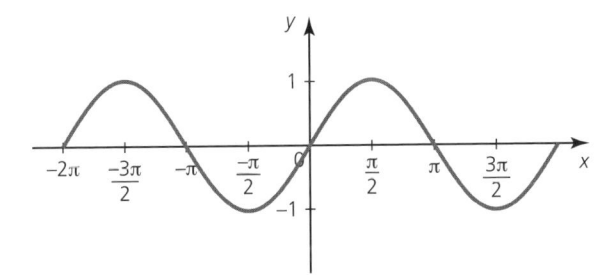

(iii) $y = |\sin|x||$

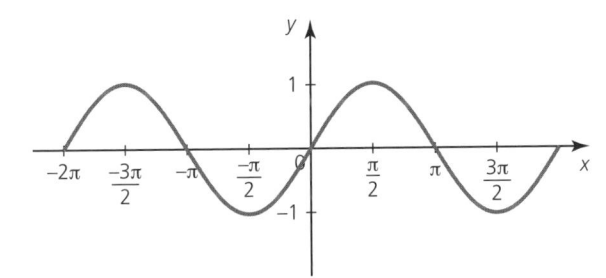

4 Solve the following equations.

(i) $|x-1| = 2$

(ii) $|2x+3| = 7$

(iii) $7 - |3+4x| = 2$

(iv) $|x+4| = |x-2|$

(v) $|2x-1| = |x+3|$

(vi) $|3x+4| = |4-x|$

5 Solve the inequalities.

(i) $|x+2| < 5$

(ii) $|3x-1| \geqslant 8$

(iii) $|x+1| \leqslant |x-2|$

(iv) $|2x-3| > |x+3|$

(v) $|4x| \leqslant |3-x|$

(vi) $3|x+2| < 1-x$

6 (i) Solve the inequality $|2x + 1| \leq |x - 3|$.

(ii) Given that x satisfies the inequality $|2x + 1| \leq |x - 3|$, find the greatest possible value of $|x + 2|$.

OCR Core Mathematics 3 4723/01 June 2010 Q5

7 Solve the equation $|3x + 4a| = 5a$, where a is a positive constant.

OCR Core Mathematics 3 4723/01 January 2011 Q1

8 Given that a is a positive constant, solve the inequality $|x - 2a| < |x + a|$.

Stretch and challenge

1 When the polynomial $f(x) = x^3 - ax^2 + 12x + b$ is divided by $g(x) = x + 5$ the quotient is $x^2 + 10x + c$ and the remainder is 150.

Find the values of a, b and c.

2 The solution to the inequality $|ax - b| \leqslant |cx - d|$ is $0 \leqslant x \leqslant 2$, where a, b, c and d are constants.

Find three different sets of values for a, b, c and d that make the inequality true.

3 Solve $\left|\dfrac{2x+1}{x-1}\right| \leqslant 2$.

4 The formula for the roots of a general quadratic equation $ax^2 + bx + c = 0$ is well known as $x = \dfrac{-b \pm \sqrt{b^2 - 4ac}}{2a}$.

A similar formula for the roots of a general cubic polynomial $ax^3 + bx^2 + cx + d = 0$ is more elusive!

(i) Given that one root of the cubic is $x = r$, show that
$$\dfrac{ax^3 + bx^2 + cx + d}{x - r} = ax^2 + (b + ar)x + (c + br + ar^2).$$

(ii) Hence solve the quadratic factor of the polynomial to obtain a formula for the other two roots of the cubic in terms of a, b, c and r.

5 The numbers a, b and c satisfy the following three equations.

$$a+b+c=5 \qquad a^2+b^2+c^2=9 \qquad \frac{1}{a}+\frac{1}{b}+\frac{1}{c}=2.$$

(i) Find the value of $ab+ac+bc$ and abc.

(ii) Show that a, b and c are the roots of the equation $x^3-5x^2+8x-4=0$.

(iii) Find the values of a, b and c.

6 A polynomial f(x) of degree 45 has a remainder of 4 when divided by $x-1$, a remainder of 7 when divided by $x-2$ and a remainder of 25 when divided by $x-4$.

Find the remainder when f(x) is divided by $(x-1)(x-2)(x-4)$.

Exam focus

1 The polynomial p(x) is defined by $p(x) = ax^3 - x^2 + 8x + a - 1$, where a is a constant.

    (i) Given that $x + 2$ is a factor of p(x), find the value of a. [2]

    (ii) When a has this value:

        (a) factorise p(x) completely [3]

        (b) find the remainder when p(x) is divided by $x - 3$. [2]

2 The polynomial $ax^3 - 8x^2 - 2x - a$ is denoted by f(x). When f(x) is divided by $x - 2$ the remainder is -15.

    (i) Find the value of a. [3]

    (ii) When a has this value find the quadratic factor of f(x). [3]

P2 1 Algebra

3 Solve the inequality $4|x - 1| \leq |x + 2|$. [4]

4 (i) Find the quotient when the polynomial $6x^3 + 13x^2 - 14x + 10$ is divided by $3x^2 + 8x - 3$ and show that the remainder is 7. [3]

(ii) Hence, or otherwise, factorise the polynomial $6x^3 + 13x^2 - 14x + 3$. [2]

5 Solve $|a - x| < 2|x + 3a|$. [4]

2 Logarithms and exponentials

Logarithms and exponential functions

EXERCISE 2.1

1. Write an equivalent logarithm statement for each of these.

   (i) $2^4 = 16$ (ii) $3^3 = 27$ (iii) $4^{-2} = \frac{1}{16}$

2. Write an equivalent index statement for each of these.

   (i) $\log_2 \frac{1}{2} = -1$ (ii) $\log_3 9 = 2$ (iii) $\log_4 2 = \frac{1}{2}$

3. Find the value of each of the following (without using a calculator).

   (i) $\log_2 8$ (ii) $\log_6 \frac{1}{216}$ (iii) $\log_{\frac{1}{3}} 27$

   (iv) $\log_{10} 100$ (v) $\log_{10} 0.1$ (vi) $\log_7 1$

   (vii) $\log_{16} 4$ (viii) $\log_5 5\sqrt{5}$ (ix) $\log_{30} 30$

   (x) $\log_{\sqrt{2}} 4$ (xi) $\log_8 \sqrt{2}$ (xii) $\log_9 \sqrt{3}$

   (xiii) $\dfrac{\log 8}{\log 2}$ (xiv) $\dfrac{\log 9}{\log 27}$ (xv) $\dfrac{\log 0.2}{\log 25}$

4. Find the value of the unknown in each equation.

   (i) $\log_2 a = 5$ (ii) $\log_3 b = -2$ (iii) $\log_c 27 = 3$

   (iv) $\log_d 8 = \frac{1}{2}$ (v) $\log_5 \frac{1}{25} = e$ (vi) $\log_{36} 6 = f$

5 Evaluate these logarithms.

(i) $\log_a a$
(ii) $\log_b b^2$
(iii) $\log_c 1$

(iv) $\log_d \sqrt[3]{d}$
(v) $\log_e \frac{1}{e}$
(vi) $\log_f \frac{1}{\sqrt{f}}$

6 Write each expression as the logarithm of a single number.

(i) $\log 5 + \log 4$
(ii) $\log 14 - \log 2$
(iii) $3\log 4$

(iv) $\frac{1}{2}\log 36$
(v) $2\log 3 + 3\log 2$
(vi) $\frac{1}{2}\log 100 - 2\log 5$

(vii) $\log 8 - \log 2 + \log 5$
(viii) $2\log 5 + \frac{1}{3}\log 64 - \frac{1}{2}\log 121$

7 If $a = \log 2$ and $b = \log 3$, write the following in terms of a and b.

(i) $\log 6$
(ii) $\log 12$
(iii) $\log \frac{9}{4}$

(iv) $\log 8$
(v) $\log 0.\dot{2}$
(vi) $\log \frac{4}{\sqrt{3}}$

8 Write the following equations without logarithms.

(i) $\log_{10} A = 2\log_{10} b + 1$

(ii) $2\log_5 D = \log_5(E-1) + \log_5 3$

9 Solve these equations.

(i) $2^x = 12$

(ii) $3^{x+1} = 24$

(iii) $6^{2x} = 4^{x-1}$

(iv) $2^{x-1} 3^x = 16$

(v) $\log_{10} x + 3 = \log_{10}(x+3)$

(vi) $\log_5(x-1) = \log_5 x - 1$

(vii) $\log_3(x+4) - \log_3(x-4) = 1$

(viii) $\log_2(x-1) = \log_2 x - 1$

(ix) $3 - 4^x = \dfrac{2}{4^x}$

(x) $7^{x+2} = 7^x + 7^2$

10 Solve the inequalities.

(i) $4^{x+2} - 2 < 18$

(ii) $|3^x - 2| \geq 1$

11 Find the values of a and b from the graphs below.

(i) $y = a(2^x) + b$

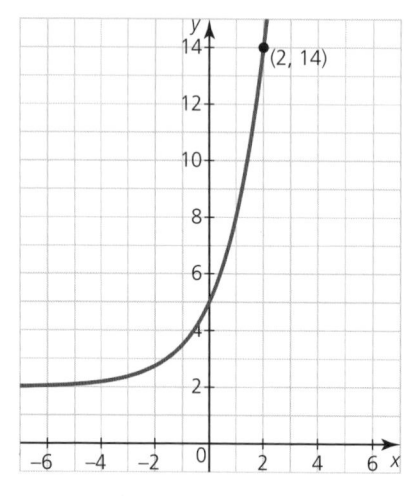

(ii) $y = \log_{10}(x+a) + b$

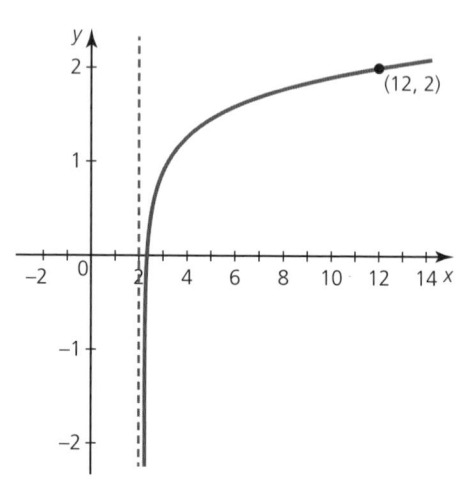

$a = $ _____ $b = $ _____ $a = $ _____ $b = $ _____

12 The magnitude (M) of an earthquake is measured on the Richter scale using the formula $M = \log_{10} \dfrac{I}{S}$, where I is the intensity of the earthquake and S is the intensity of a 'standard' earthquake.

In 2010 an earthquake in Christchurch, New Zealand registered 7.1 on the Richter scale and in 1985 Mexico City experienced an 8.3 magnitude earthquake.

(i) How many times greater was the intensity of the Mexico City earthquake than the New Zealand earthquake? Give your answer to the nearest whole number.

(ii) Find the magnitude of an earthquake that would be half the intensity of the earthquake in Mexico City.

(iii) Find the magnitude of an earthquake that would be double the intensity of the earthquake in Christchurch.

13 The loudness of a sound (L) is measured in decibels (dB) according to the formula $L = 20\log_{10}\left(\dfrac{P}{P_o}\right)$, where P is the power (or intensity) of the sound and P_o is a fixed reference power.

A rock band registers at 110 dB and a plane taking off is 125 dB.

How many times greater is the intensity of the sound of the plane compared to the rock band?

14 Solve simultaneously.
$$2\log_{10} x + \log_{10} y = 2$$
$$xy^2 = 80$$

Modelling curves

EXERCISE 2.2

1 (i) Show that the equation $y = kx^p$ can be written in the form $\log y = p \log x + \log k$.

(ii) Hence state the gradient and y intercept of the straight line on the graph of $\log y$ against $\log x$.

2 The equation of the line of best fit for a set of data on the graph of $\log_{10} y$ against x is $\log_{10} y = 0.7x + 3.5$.

Find a suitable model for the data in the form $y = A(b)^x$.

3 The average weight loss (W kg) of a large group of people on a diet is measured after 1, 2, 5 and 10 months (m) on the diet.

It is proposed that the average weight loss after m months on the diet can be modelled by an equation of the form $W = Am^b$, where A and b are constants.

(i) Complete the table.

m	W	$\log_{10} m$	$\log_{10} W$
1	8.00		
2	5.66		
5	3.58		
10	2.53		

(ii) The graph of $\log_{10} W$ against $\log_{10} m$ is shown on the axes.

Use the graph to determine the values of the constants A and b.

(iii) Based on this model:

(a) calculate the average weight loss at 12 months

(b) find when the average weight loss will be less than 1 kg.

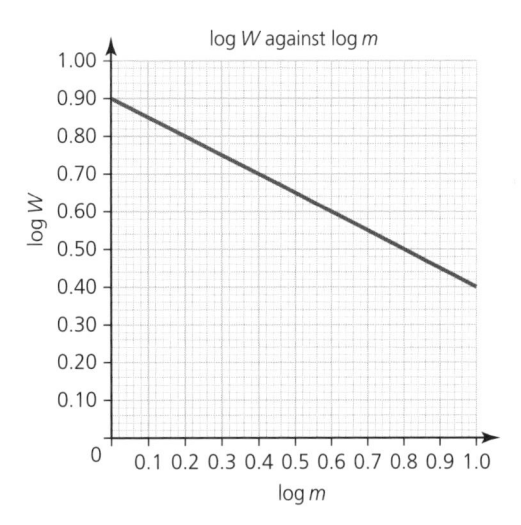

4 The average length (L cm) of male babies born in Karachi is measured at regular intervals.

Let t be the age in months of the babies.

A doctor proposed that the data can be modeled with an equation of the form $L = Ab^t$, where A and b are constants.

t	L	$\log_{10} L$
0	50	
5	57	
10	66	
15	77	
20	87	
36	128	

(i) Complete the table.

(ii) Show that the equation $L = Ab^t$ can be reduced to linear form by taking logarithms of both sides.

(iii) The graph of $\log_{10} L$ against t is shown. Use the graph to determine the values of A and b.

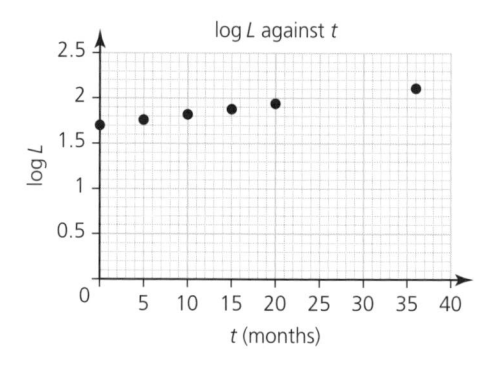

log L against t

(iv) What does the model predict the height of an average 18-year-old man will be?

Is your answer reasonable? Why or why not?

5 The line with gradient –2 and y intercept 1 fits the points on the graph of $\log_{10} y$ against $\log_{10} x$.

The equation $y = Ax^b$ models the relationship between x and y.

(i) Find the values of A and b.

(ii) Find the value of y when $x = 20$.

6 The variables x and y satisfy the equation $y = kb^{-x}$, where k and b are constants.

The graph of $\log_{10} y$ against x is a straight line passing through the points (1.3, 3.4) and (4.2, 0.7).

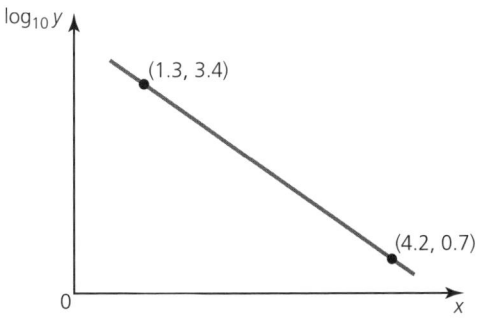

Find the values of k and b correct to 2 decimal places.

7 A golfer wants to model the distance she hits the ball (D) as a function of the loft in degrees (L) of the club.

To do this she hits 100 golf balls and finds the average distance the ball travels for each club.

The graph of $\log_{10} D$ against L is shown.

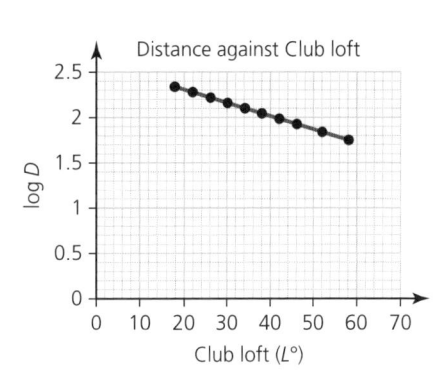

(i) Two models are proposed initially:

Model A: $D = kL^p$ and

Model B: $D = k(p)^L$

where k and p are constants.

Based on the graph, which model is correct? State a reason.

(ii) Find the values of k and p.

(iii) Using this model find the loft that would be needed to hit the ball 120 m on average.

(iv) Discuss the limitations of the model.

8 The variables x and y are related by the equation $y^4 = mx^n$, where m and n are constants. The graph of $\log_{10} y$ against $\log_{10} x$ is a straight line passing through (0, 0.25) and (4, 5).

Find the values of m and n.

The natural logarithm function

EXERCISE 2.3

1 Solve the following equations, giving your answer exactly.

   (i) $e^{2x+1} = 4$

   (ii) $\ln 2x = \frac{1}{2}\ln 16 + \frac{2}{3}\ln 8$

   (iii) $\ln(x+1) + 1 = 3$

   (iv) $e^{2x} + e^x = 30$

   (v) $\ln x - \ln 4 = \ln(x-4)$

   (vi) $3e^{-x} + 1 = 2e^x$

   (vii) $\ln x = \ln(x+1) + 1$

   (viii) $e^{3x} - e^{2x} = 2e^x$

2 Write these equations without logarithms.

   (i) $\ln A = 3\ln B + 2\ln 3$

   (ii) $\frac{1}{2}\ln P - \ln Q = 3\ln R + 1$

3 Simplify.

(i) $\ln(e^{x+y})^2$
(ii) $e^{2\ln x + 3\ln y}$
(iii) $2\ln\sqrt{e^{x-y}}$

4 The variables x and y satisfy the equation $y = Ax^b$.

Given that the graph of $\ln y$ against $\ln x$ is a straight line passing through $(1.2, 0.5)$ and $(8.1, 12.3)$, find the values of the constants A and b.

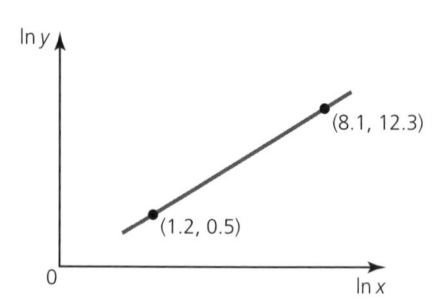

5 The number of bacteria in a colony, N, can be modelled by the equation $N = 1000e^{0.4t}$, where t is time in hours since measurements were started.

(i) What is the initial size of the colony?

(ii) Find the number of bacteria after 5 hours.

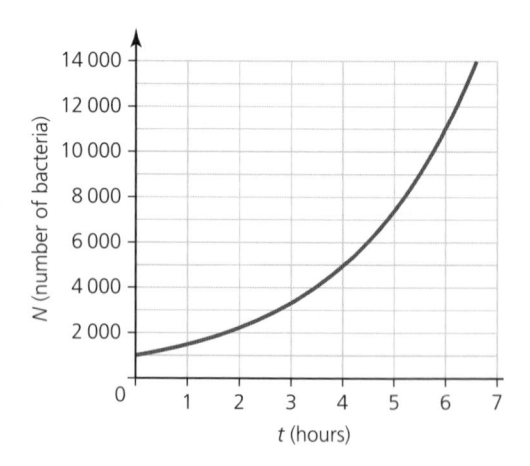

(iii) Calculate how long (to the nearest minute) it took for the bacteria to double in number.

(iv) Find how many hours it would take for the number of bacteria in the colony to pass 1 million.

6 Find the x co-ordinate of the point of intersection of the two curves $y = e^{x-1}$ and $y = e^{-x}$.

7 The diagram shows the curve $y = e^{kx} - a$, where k and a are constants.

  (i) Sketch the curve $y = |e^{kx} - a|$ on the axes.

  (ii) Given that the curve $y = |e^{kx} - a|$ passes through the points (0, 13) and (ln 3, 13), find the values of k and a.

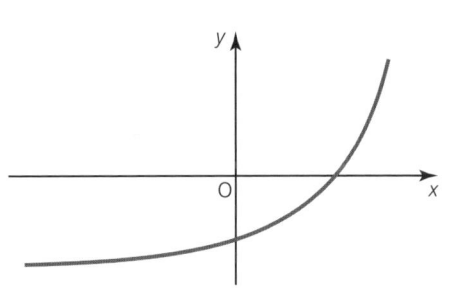

OCR Core Mathematics 3 4723/01 January 2009 Q7

8 It is given that $p = e^{280}$ and $q = e^{300}$.

  (i) Use logarithm properties to show that $\ln\left(\dfrac{ep^2}{q}\right) = 261$.

  (ii) Find the smallest integer n which satisfies the inequality $5^n > pq$.

OCR Core Mathematics 3 4723/01 June 2012 Q2

Stretch and challenge

1 Evaluate these expressions.

(i) $\log_4 \sqrt[5]{\frac{32}{1024}}$

(ii) $\log_2 \left[\frac{\sqrt{256}\left(\frac{1}{8}\right)^6}{32\left(\frac{1}{2}\right)^3} \right]$

(iii) $e^{\ln e^{\ln \pi}}$

(iv) $36^{\frac{1}{2} - \log_6 \sqrt{3}}$

2 For the following equations state whether they are:

Always true   Sometimes true   Never true

Explain your reasoning.

(i) $\log_a b = \log_b a$

(ii) $\log_{\frac{1}{a}} a = 1$

(iii) $\log_a a = 0$

(iv) $\log_a (\log_a a) = 1$

(v) $\frac{\log a}{\log b} = \log a - \log b$

(vi) $\log_a \log_b \log_c c = 0$

(vii) $\log_a x + \log_b x = \log_{ab} x$

3 Solve the equation $3(2^x) - 4^x = 2$.

4 Solve $\log_3 x + \log_4 x - 1 = \log_5 x$.

5 Show that if $3\log_x y + 3\log_y x = 10$ then $y = x^3$ or $x = y^3$.

6 If $2\log(x - 2y) = \log x + \log y$ find the possible values of $\dfrac{x}{y}$.

7 The average speeds of cars (S km/h) along a stretch of highway have been measured over many years.

Year	1990	1995	2000	2005	2010
Average speed, S	103.2	104.5	106.3	108.5	112.4

A model of the average speeds is given by $S = 100 + ab^t$, where t is years from 1990 and a and b are constants.

(i) By drawing a suitable graph for $0 \leq t \leq 20$, find the values of a and b.

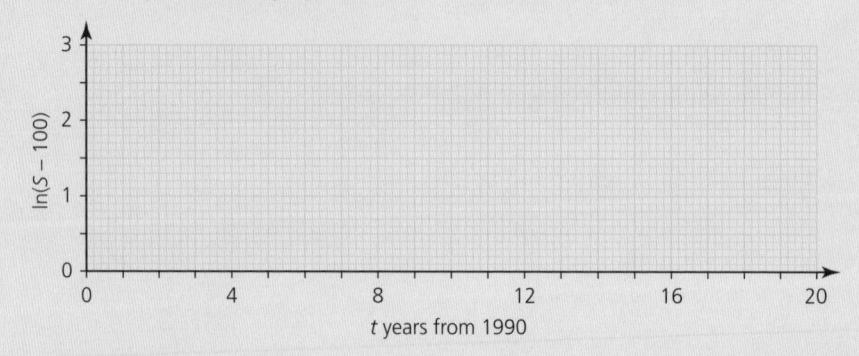

(ii) According to the model, what will the average speed of cars be on the highway in 2030?

(iii) Discuss any limitations of the model.

8 The variables x and y are related so that when xy is plotted against x^2, the result is a straight line passing through the points $(4, 6)$ and $(9, 21)$ as shown.

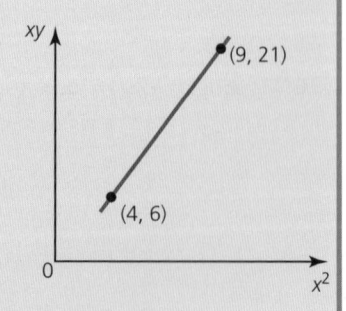

(i) Find the value of y when $x = 6$.

(ii) Find the two possible values of x when $y = 7$.

9 Solve the equation $3^{\log_{10} x^2} = 2(3^{1+\log_{10} x}) + 27$.

10 Find all **real** solutions of $\dfrac{ae^x}{2e^x - 1} < 1$, where a is a positive constant.

NZQA Scholarship Calculus 2011 Q1a

Exam focus

1 The variables x and y are related by the equation $y = Ab^x$ where A and b are constants. The graph of $\ln y$ against x is a straight line passing through the points shown.

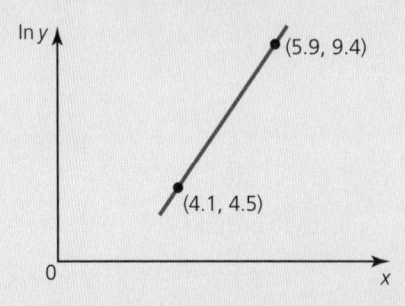

$\ln y$

(5.9, 9.4)

(4.1, 4.5)

0 x

Find the values of A and b to 3 significant figures. [6]

2 Solve the equation $6^x = 6^{x-1} + 6$ giving your answer correct to 3 significant figures. [4]

3 (i) Show that the equation $\log_4(x-4) = 2 - \log_4 x$ can be written as a quadratic equation in x. [3]

(ii) Hence solve the equation $\log_4(x-4) = 2 - \log_4 x$ giving your answer to 2 decimal places. [2]

4 Solve the inequality $|8 - 3^x| = 15$. [3]

5 Solve the equation $\ln(3x+2) = 2\ln x + \ln 2$. [4]

6 Solve the equation $5^{x-1} = 7^{2x-1}$. [4]

P2 3 Trigonometry

Reciprocal trigonometrical functions

EXERCISE 3.1

1. The graphs of $y = \sin x$, $y = \cos x$ and $y = \tan x$ for $-2\pi \leq x \leq 2\pi$ are shown below.

   Use these to sketch the following graphs.

   (i) $y = \operatorname{cosec} x = \dfrac{1}{\sin x}$

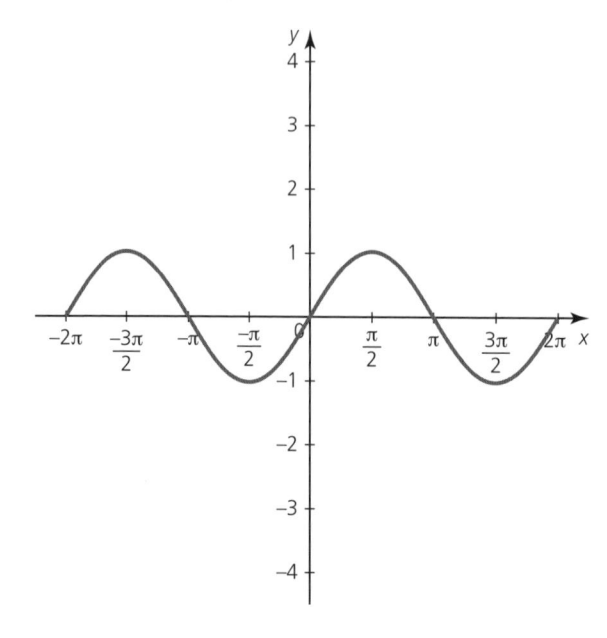

(ii) $y = \sec x = \dfrac{1}{\cos x}$

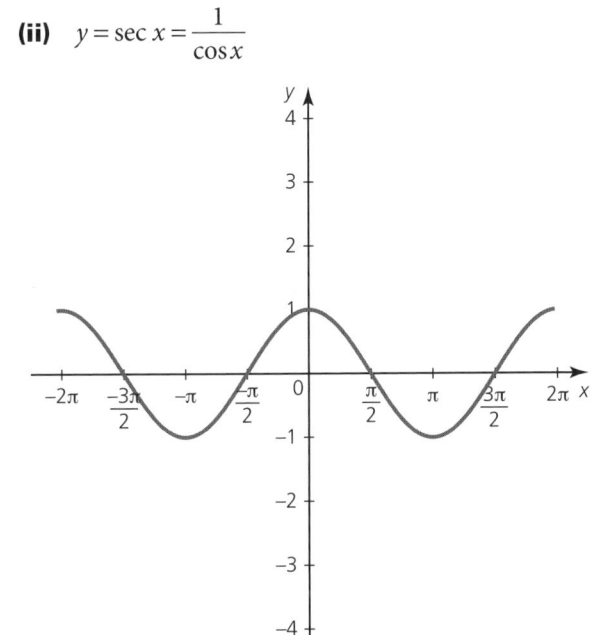

(iii) $y = \cot x = \dfrac{1}{\tan x} = \dfrac{\cos x}{\sin x}$

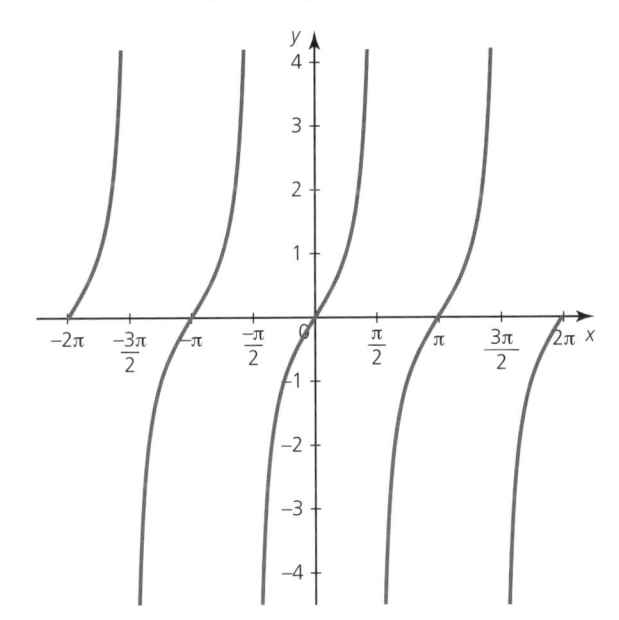

2 Find the exact value (without using a calculator) of the following.

(i) cosec 150° = _____

(ii) $\sec \dfrac{\pi}{4} =$ _____

(iii) cot 300° = _____

(iv) $\operatorname{cosec} \dfrac{4\pi}{3} =$ _____

(v) sec 120° = _____

(vi) $\cot \dfrac{3\pi}{4} =$ _____

3 Starting with the identity $\sin^2\theta + \cos^2\theta \equiv 1$, show the following.

   (i) $\tan^2\theta + 1 \equiv \sec^2\theta$ **(ii)** $1 + \cot^2\theta \equiv \text{cosec}^2\theta$

4 Eliminate θ from these equations.

   (i) $x = 2\,\text{cosec}\,\theta \quad y = 3\cot\theta$ **(ii)** $x = \sin\theta - \cos\theta \quad y = \sin\theta + \cos\theta$

5 Solve the following equations over the given domains.

   (i) $\text{cosec}\,\theta = 4$ for $0° \leqslant \theta \leqslant 360°$ **(ii)** $\sec\tfrac{1}{2}\theta = 4$ for $0° \leqslant \theta \leqslant 360°$

   (iii) $\tan\beta = 7\cot\beta$ for $0° < \beta < 180°$ **(iv)** $\tan x \tan 2x = 4$ for $0° \leqslant x \leqslant 180°$

   (v) $\tan\theta + \cot\theta = -4$ for $-\ \leqslant \theta \leqslant$

6 Given that $\sec\theta = 3$ and $0° \leq \theta \leq 90°$, find the exact values of the following.

(i) $\cos\theta$ (ii) $\sin\theta$

(iii) $\csc\theta$ (iv) $\cot\theta$

7 Prove the following identities.

(i) $\dfrac{1}{\tan\theta + \cot\theta} \equiv \sin\theta\cos\theta$ (ii) $\sec^2\theta + \csc^2\theta \equiv \sec^2\theta \csc^2\theta$

(iii) $\sec^4\theta - \tan^4\theta \equiv \sec^2\theta + \tan^2\theta$ (iv) $(\tan\theta - \sin\theta)^2 + (1 - \cos\theta)^2 = (1 - \sec\theta)^2$

(v) $(\csc^2\theta - 1)(\tan^2\theta + 1) \equiv \csc^2\theta$ (vi) $\dfrac{\cos\theta}{1 - \tan\theta} - \dfrac{\sin\theta}{1 - \cot\theta} \equiv \dfrac{1}{\cos\theta - \sin\theta}$

8 $m = \dfrac{1+\cos\theta}{\sin\theta}$

(i) Show that $\dfrac{1}{m} = \dfrac{1-\cos\theta}{\sin\theta}$.

(ii) Find an expression for $\cos\theta$ in terms of m only.

Compound-angle formulae

EXERCISE 3.2

1 Find the exact value of the following.

(i) $\cos 75° = \cos(45° + 30°)$

(ii) $\sin 15° = \sin(60° - 45°)$

(iii) $\tan 105°$

(iv) $\sec 15°$

2 Simplify each of these as much as possible.

(i) $\sin(\theta - 30°)$

(ii) $\cos\left(\dfrac{\pi}{4} - \theta\right)$

(iii) $\tan(\theta + 60°)$

(iv) $\operatorname{cosec}(2\theta + 120°)$

(v) $\dfrac{\cos A + \cos(-A)}{\sin A - \sin(-A)}$ (vi) $\dfrac{\cos(30°+A)-\cos(30°-A)}{\sin(30°+A)-\sin(30°-A)}$

3 Write each of the following expressions in the form $\sin(A \pm B)$ or $\cos(A \pm B)$.

  (i) $\sin\theta \cos 2\beta + \sin 2\beta \cos\theta$ (ii) $\cos 3\theta \cos\theta + \sin 3\theta \sin\theta$

  (iii) $\sin\dfrac{4}{3}\cos\dfrac{7}{6} - \sin\dfrac{7}{6}\cos\dfrac{4}{3}$ (iv) $\cos 280° \cos 20° - \sin 280° \sin 20°$

4 Use the diagram to find the exact value of $\sin(x+y)$.

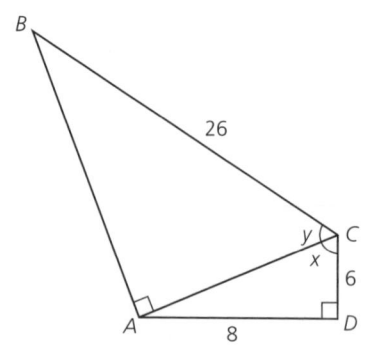

5 The angles P and Q are both acute with $\cos P = \frac{2}{5}$ and $\tan Q = \frac{7}{3}$.

Find the exact value of the following.

(i) $\cos(P - Q)$ **(ii)** $\sin(P + Q)$

6 Given that A is an obtuse angle with $\sin A = \frac{1}{3}$ and B is an acute angle with $\sec B = 4$, find the **exact** value of each of these.

(i) $\sin(A - B)$ **(ii)** $\cot(B - A)$

7 Solve the equations.

(i) $\sin(45° - \theta) = \cos(30° + \theta)$ for $-180° \leqslant \theta \leqslant 180°$

(ii) $\tan\left(\dfrac{\pi}{3}+\theta\right)=2\tan\left(\dfrac{\pi}{6}-\theta\right)$ for $0\leqslant\theta\leqslant\pi$

(iii) $\tan(\theta+45°)=1-2\tan\theta$ for $0°\leqslant\theta\leqslant90°$

8 Prove $\cos(A+B)\cos(A-B)\equiv\cos^2 A-\sin^2 B$.

9 A and B are acute angles with $\tan A=\tfrac{1}{2}$ and $\tan B=\tfrac{2}{3}$.

Find the exact value of the following.

(i) $\tan(A+B)$ (ii) $\cos(A-B)$

10 A is an acute angle and B is an obtuse angle such that $\tan A = \frac{1}{3}$ and $\tan(A - B) = 5$.

  (i) Find $\tan B$.

  (ii) Hence show that the exact value of $\sin(A + B)$ is $\dfrac{17}{\sqrt{650}}$.

Double-angle formulae

EXERCISE 3.3

1 Use the double-angle formulae to find the exact value of the following.

(i) $\sin\dfrac{2}{3}$ (ii) $\cos\dfrac{2}{3}$ (iii) $\tan\dfrac{2}{3}$

2 Given θ from the triangle shown, find the exact value of each of these.

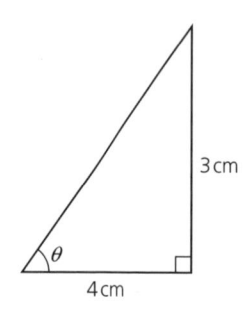

(i) $\sin 2\theta$ (ii) $\cos 2\theta$

3 Given that $\sin\theta = -\dfrac{2}{3}$ and $\dfrac{3}{2} \leqslant \theta \leqslant 2$, find the exact value of the following.

(i) $\sin 2\theta$ (ii) $\cos 2\theta$ (iii) $\sin 4\theta$

4 Find an expression for $\cos 4\theta$ in terms of $\cos\theta$.

5 Starting with $\sin 3\theta = \sin(2\theta + \theta)$, find an expression for $\sin 3\theta$ in terms of $\sin\theta$.

6 (i) If $5 + 4\sec^2\theta = 12\tan\theta$, find the exact value of $\tan\theta$.

 (ii) Hence find the exact value of these expressions.

 (a) $\tan(\theta + 45°)$ **(b)** $\tan 2\theta$

7 Solve the equations.

(i) $\sin 2\theta = \sin\theta$ for $0 \leq \theta \leq 2$

(ii) $\cos 2\theta - 5\sin\theta = 3$ for $0° \leq \theta \leq 360°$

(iii) $3\tan 2\theta + 2\tan\theta = 0$ for $-180° < \theta < 180°$

(iv) $2\cos 2\theta = 1 + \cos\theta$ for $0° < \theta < 360°$

(v) $\dfrac{1-\sin\theta - \cos 2\theta}{\cos\theta - \sin 2\theta} = 1$ for $0° \leq \theta \leq 360°$

(vi) $\tan 2\theta = \dfrac{1}{1+\tan\theta}$ for $-\quad \leq \theta \leq$

8 Prove the identities.

   (i) $\sin(45° + \theta)\sin(45° - \theta) \equiv \frac{1}{2}\cos 2\theta$

   (ii) $\dfrac{2\sin\theta\cos\theta}{\cos^4\theta - \sin^4\theta} \equiv \tan 2\theta$

   (iii) $\dfrac{\sin 3\theta}{\sin\theta} - \dfrac{\cos 3\theta}{\cos\theta} = 2$

9 (i) Prove the identity $\cot\theta + \tan\theta \equiv 2\operatorname{cosec} 2\theta$.

   (ii) Hence solve the equation $\cot\theta + \tan\theta = 8$ for $0 \leqslant \theta \leqslant 2$.

10 (i) Prove the identity $\sin 2\theta + 2\tan 2\theta \sin^2\theta \equiv \tan 2\theta$.

(ii) Hence solve $\sin 2\theta + 2\tan 2\theta \sin^2\theta = 7$ for $0° \leqslant \theta \leqslant 180°$

11 If $\sin 25° = k$, where k is a positive constant, express the following in terms of k.

(i) $\sin 50°$ **(ii)** $\cos 50°$ **(iii)** $\tan 155°$

12 The value of tan 10° is denoted by p. Find, in terms of p, the value of

(i) $\tan 55°$

(ii) $\tan 5°$

(iii) $\tan\theta$, where θ satisfies the equation $3\sin(\theta + 10°) = 7\cos(\theta - 10°)$.

OCR Core Mathematics 3 4723/01 January 2010 Q9

13 By writing $\tan 3x = \tan(2x + x)$, find an expression for $\tan 3x$ in terms of $\tan x$.

14 It is given that θ is the acute angle such that $\sec\theta \sin\theta = 36\cot\theta$.

  (i) Show that $\tan\theta = 6$.

  (ii) Hence, using an appropriate formula in each case, find the exact value of

  (a) $\tan(\theta - 45°)$ **(b)** $\tan 2\theta$.

OCR Core Mathematics 3 4723/01 June 2012 Q3

The forms $r\cos(\theta \pm \alpha)$, $r\sin(\theta \pm \alpha)$

EXERCISE 3.4

1 Write these expressions in the form given where $r > 0$ and $0° < \alpha < 90°$.

(i) $\sin\theta - 3\cos\theta$
in the form $r\sin(\theta - \alpha)$

(ii) $12\cos\theta + 5\sin\theta$
in the form $r\cos(\theta - \alpha)$

(iii) $6\sin\theta + 8\cos\theta$
in the form $r\sin(\theta + \alpha)$

(iv) $7\cos\theta - 24\sin\theta$
in the form $r\cos(\theta + \alpha)$

2 (i) Express $2\cos\theta + \sin\theta$ in the form $r\cos(\theta - \alpha)$ where $r > 0$ and $0° < \alpha < 90°$.

(ii) Hence solve $2\cos\theta + \sin\theta = 1$ for $0° \leqslant \theta \leqslant 360°$.

(iii) Sketch the graph of $y = 2\cos\theta + \sin\theta$ on the axes.

(iv) Find the greatest and least value of $2\cos\theta + \sin\theta + 5$ as θ varies.

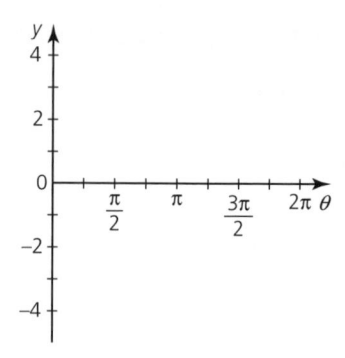

3 (i) Express $4\sin\theta - 3\cos\theta$ in the form $R\sin(\theta - \alpha)$ where $r > 0$ and $0° < \alpha < 90°$.

(ii) Hence solve $4\sin\theta - 3\cos\theta = 3$ for $0° \leqslant \theta \leqslant 360°$.

(iii) Find the maximum and minimum value of $4\sin\theta - 3\cos\theta + 6$ as θ varies and give the smallest positive value of θ where the maximum and minimum occurs.

4 The expression $T(\theta)$ is defined for θ in degrees by

$T(\theta) = 3\cos(\theta - 60°) + 2\cos(\theta + 60°)$.

(i) Express $T(\theta)$ in the form $A\sin\theta + B\cos\theta$, giving the exact values of the constants A and B.

(ii) Hence express $T(\theta)$ in the form $R\sin(\theta + \alpha)$, where $R > 0$ and $0° < \alpha < 90°$.

(iii) Find the smallest positive value of θ such that $T(\theta) + 1 = 0$.

5 (i) Express $3\sin\theta + 4\cos\theta$ in the form $R\sin(\theta + \alpha)$, where $R > 0$ and $0° < \alpha < 90°$.

(ii) Hence

  (a) solve the equation $3\sin\theta + 4\cos\theta + 1 = 0$, giving all solutions in the interval $-180° < \theta < 180°$

  (b) find the values of the positive constants k and c such that $-37 \leqslant k(3\sin\theta + 4\cos\theta) + c \leqslant 43$ for all values of θ.

OCR Core Mathematics 3 4723/01 June 2012 Q8

6 (i) Express $\sqrt{2}\cos\theta + \sqrt{7}\sin\theta$ in the form $R\cos(\theta - \alpha)$, where $R > 0$ and $0° < \alpha < 90°$. Give the value of α correct to 2 decimal places.

(ii) Hence, in each of the following cases, find the smallest positive angle θ which satisfies the equation.

(a) $\sqrt{2}\cos\theta + \sqrt{7}\sin\theta = -1$

(b) $\sqrt{2}\cos\frac{1}{2}\theta + \sqrt{7}\sin\frac{1}{2}\theta = 2$

Stretch and challenge

1 A projectile is fired from a sloping hill that makes an angle A with the horizontal. It is fired with velocity V m/s at an angle B to the hill as shown.

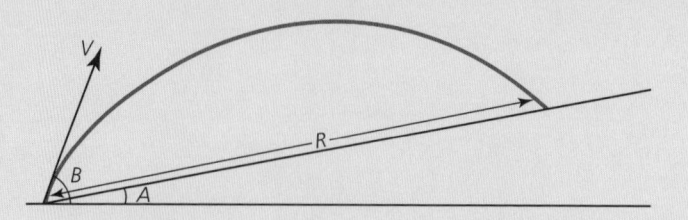

The range, R, that the projectile can travel is given by

$$R = \frac{2V^2 \sin B}{g \cos^2 A} \cos(A + B)$$

where g = acceleration due to gravity = 10 m/s².

   (i) Express R as a function of B given that $A = \frac{\pi}{4}$.

   (ii) If the projectile is fired at a speed of 40 m/s, find the angle B it should be fired at to hit a target with a range of 150 m.

2 Square 12 cm × 12 cm floor tiles in the design shown are laid at an angle θ to the vertical.

The tiles are laid in strips that are 15 cm wide.

(i) Find the value of θ and hence find the overlap, w cm between the tiles.

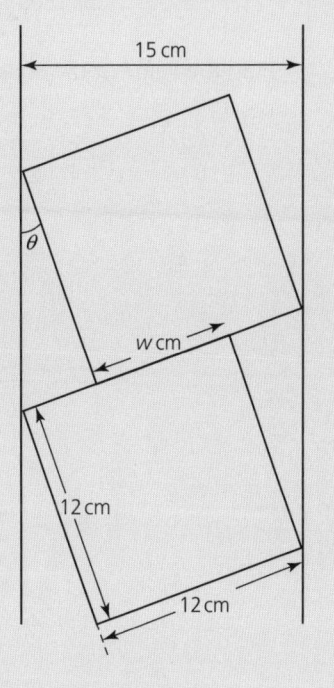

(ii) Another tiler wants to make the strips wider than 15 cm.

Find the maximum possible width of the strips that will work with the 12 cm × 12 cm tiles.

3 For all values of x for which the terms are defined, it is given that
$$\tan x - \tan \tfrac{1}{8}x = \frac{\sin kx}{\cos x \cos \tfrac{1}{8}x}.$$

Find the value of the constant k.

4 If $\sin A + \cos A = 1.5$, find the value of $\sin^3 A + \cos^3 A$.

5 If $\cos\theta = 0.1$ and $0 \leq \theta \leq \dfrac{\pi}{2}$, find the value of $\log_{10}(\tan\theta) - \log_{10}(\sin\theta)$.

6 An amusement park has a giant double Ferris wheel as shown in the diagram.

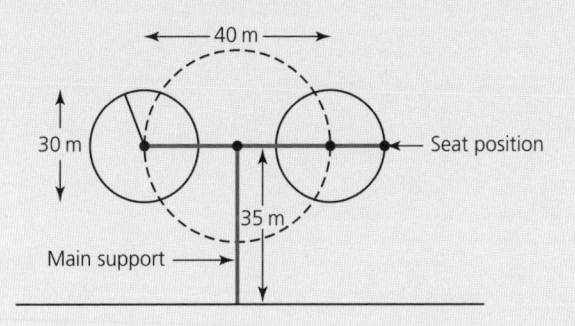

The double Ferris wheel has a rotating arm 40 metres long attached at its centre to a main support 35 metres above the ground. At each end of the rotating arm is attached a Ferris wheel measuring 30 metres in diameter, as shown in the diagram. The rotating arm takes 4 minutes to complete one full revolution, and each wheel takes 3 minutes to complete a revolution about that wheel's hub. All revolutions are anticlockwise, in a vertical plane.

At time $t = 0$ the rotating arm is parallel to the ground and your seat is at the 3-o'clock position of the rightmost wheel.

Find a formula for h(t), your height above the ground in metres, as a function of time in minutes.

NZQA Scholarship Calculus 2007 Q2b

7 A rectangular piece of paper of width 8 cm has one corner folded down so the corner rests against the opposite longer side as shown.

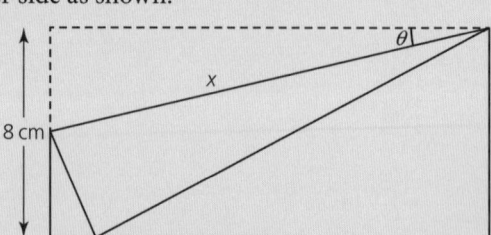

Show that $x = \dfrac{4}{\sin\theta \cos^2\theta}$.

8 Two half-angle formulae for trigonometry are given below.

$$\cos\left(\frac{\alpha}{2}\right) = \pm\sqrt{\frac{1+\cos\alpha}{2}} \qquad \sin\left(\frac{\alpha}{2}\right) = \pm\sqrt{\frac{1-\cos\alpha}{2}}$$

Given that $\tan\theta = 20\sqrt{6}$ and $0 < \theta < \frac{\pi}{2}$, find an exact value of $\tan\left(\frac{\theta}{4}\right)$.

Simplify your answer.

NZQA Scholarship Calculus 2011 Q3b

Exam focus

1. Solve the equation $\sin(x-45°)-\cos(45°-x)=1$ for $0° \leq x \leq 360°$. [4]

2. Solve the equation $2\sin(x+30°)=\cos(x-45°)$ giving all solutions in the interval $0° \leq \theta \leq 180°$. [4]

3. (i) Prove the identity $2\operatorname{cosec} 2\theta \equiv \sec\theta \operatorname{cosec}\theta$. [3]

(ii) Hence solve the equation $\sec\theta \operatorname{cosec}\theta = 4$ for $0° \leq \theta \leq 180°$. [3]

4 Solve the equation $4\operatorname{cosec}^2\theta - 7 = 4\cot\theta$ for $0° \leq \theta \leq 180°$. [4]

5 (i) Express $8\sin\theta + 15\cos\theta$ in the form $R\sin(\theta + \alpha)$ where $R > 0$ and $0° < \alpha < 90°$. [3]

(ii) Hence solve $8\sin\theta + 15\cos\theta = 14$ for $0° \leq \theta \leq 360°$. [4]

(iii) Find the range of values of the constant k such that the equation $8\sin\theta + 15\cos\theta = k$ has no solutions. [1]

6 (i) Prove the identity $\tan(\theta+60°)\tan(\theta-60°) = \dfrac{\tan^2\theta - 3}{1 - 3\tan^2\theta}$. [4]

(ii) Solve, for $0° < \theta < 180°$, the equation $\tan(\theta+60°)\tan(\theta-60°) = 4\sec^2\theta - 3$, giving your answers correct to the nearest $0.1°$. [5]

(iii) Show that, for all values of the constant k, the equation $\tan(\theta+60°)\tan(\theta-60°)=k^2$ has two roots in the interval $0° < \theta < 180°$. [3]

OCR Core Mathematics 3 4723/01 June 2007 Q9

7 (i) Express $\tan 2\alpha$ in terms of $\tan \alpha$ and hence solve, for $0° < \alpha < 180°$, the equation $\tan 2\alpha \tan \alpha = 8$. [5]

(ii) Given that β is the acute angle such that $\sin \beta = \frac{6}{7}$, find the exact value of:

(a) $\operatorname{cosec} \beta$ [1] **(ii)** $\cot^2 \beta$. [2]

OCR Core Mathematics 3 4723/01 June 2008 Q5

Differentiation

The product and quotient rules

EXERCISE 4.1

1 Differentiate the following functions.

(i) $y = (x-2)(x+3)^2$

(ii) $y = \dfrac{x^3}{x-1}$

(iii) $y = x^3(1-2x)^4$

(iv) $y = \dfrac{x+2}{3x^2}$

(v) $y = 3x^2\sqrt{1+4x}$

(vi) $y = \dfrac{2x}{\sqrt{6x-1}}$

2 Find the equation of the tangent to the curve $f(x) = 4x(x-3)^5$ at the point (4, 16).

3 Find the equation of the normal to the curve $g(x) = \dfrac{2x-1}{1-x^2}$ at the point $(2, -1)$.

4 Find the x values of the stationary points on the curve $h(x) = x^2(x+3)^3$.

5 A curve is defined by $y = \dfrac{2x-1}{x+2}$.

  (i) Find the gradient of the curve at the point where it crosses the x axis.

  (ii) Does the curve have a stationary value? Why or why not?

6 A curve is given by $f(x) = (x-1)^k (x+2)^{k+1}$, where k is a positive constant.

  (i) Two stationary points on the curve are at $x = -2$ and $x = 1$. Find the x co-ordinate of the third stationary point in terms of k.

(ii) The third stationary point occurs when $x = -\frac{1}{3}$. Calculate the value of k.

7 The graph shows the curve $y = \dfrac{x^2 - x + 2}{x + 1}$.

(i) Find $\dfrac{dy}{dx}$.

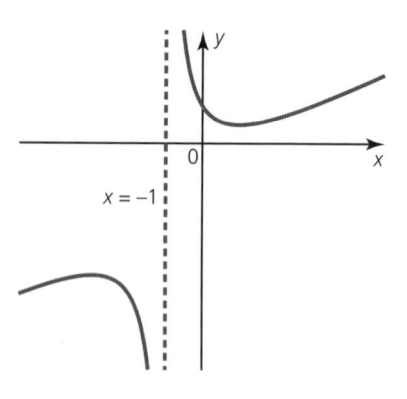

(ii) Find the co-ordinates of the stationary points on the curve.

Differentiating natural logarithms and exponentials

EXERCISE 4.2

1 Differentiate the following functions, simplifying your answers as much as possible.

(i) $y = \ln(x+4)$

(ii) $y = \ln(3x^2)$

(iii) $y = e^{1-x}$

(iv) $y = 3e^{x^3+1}$

(v) $y = 3\ln\left(\dfrac{x}{x-1}\right)$

(vi) $y = 5\ln(1+\sqrt{x})^2$

(vii) $y = xe^{x-2}$

(viii) $y = \dfrac{x^3}{e^x}$

(ix) $y = \dfrac{1+e^x}{1-e^{-x}}$

(x) $y = \sqrt{e^x}\,\ln x^2$

(xi) $y = \dfrac{e^{3x}-1}{\ln(3x-1)}$

(xii) $y = \ln(1+e^{2x})e^{5x}$

2 (i) The diagram shows the graph of $y = \dfrac{x}{\ln x}$ for $x > 1$.

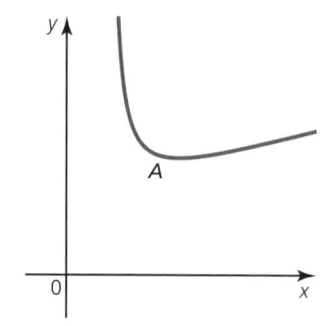

Find $\dfrac{dy}{dx}$ and $\dfrac{d^2y}{dx^2}$.

(ii) Hence:

(a) find the exact co-ordinates of the stationary point A on the curve

(b) find the exact co-ordinates of the point where the gradient is a maximum.

3 The graph of $y = x^2 e^{-x}$ is shown.

(i) Find $\dfrac{dy}{dx}$ and $\dfrac{d^2 y}{dx^2}$.

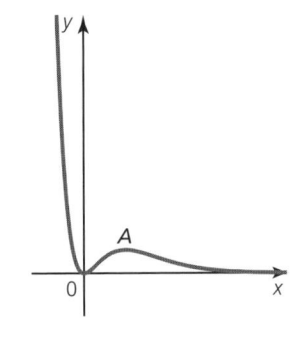

(ii) Find the exact co-ordinates of the stationary point A on the curve.

(iii) Find the x values of the two points on the curve where $\dfrac{d^2 y}{dx^2} = 0$.

What do these points represent on the graph?

4 A curve is defined by $y = \dfrac{x}{3 + 2\ln x}$.

Find the exact co-ordinates of the stationary point on the curve.

5 Find the exact co-ordinates of the stationary point on the curve $y = x^3 \ln x$ for $x > 0$.

6 The diagram shows the curve $y = 2x - x \ln x$, where $x > 0$.

The curve crosses the x axis at A, and has a stationary point at B.

The point C on the curve has x co-ordinate 1.

Lines CD and BE are drawn parallel to the y axis.

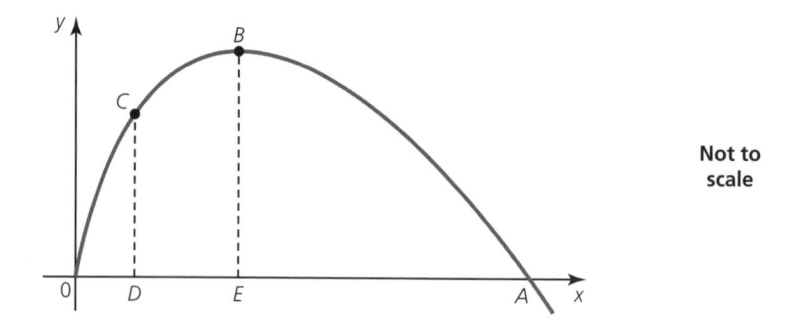

Not to scale

(i) Find the x co-ordinate of A, giving your answer in terms of e.

(ii) Find the exact co-ordinates of B.

(iii) Show that the tangents at A and C are perpendicular to each other.

OCR MEI Structured Mathematics C3 4753/1 January 2006 Q7

7 The diagram shows the graph of $y = (x+1)e^{\frac{1}{2}x}$.

(i) Find $\dfrac{dy}{dx}$.

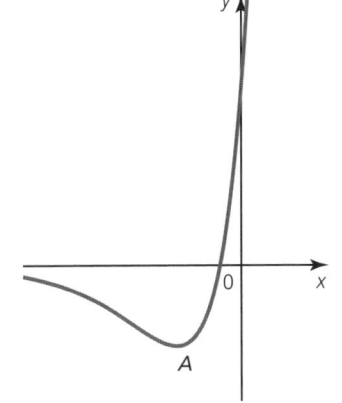

(ii) Hence find the exact co-ordinates of the minimum point A.

Differentiating trigonometrical functions

EXERCISE 4.3

1 Differentiate the following functions.

(i) $y = 3\sin 2x$

(ii) $y = \cos(1 + 3x)$

(iii) $y = \tan(x^2)$

(iv) $y = x^3 \sin 2x$

(v) $y = \dfrac{\cos 5x}{x^2}$

(vi) $y = \dfrac{\sin^3 x}{x^2}$

(vii) $y = e^{\sin x + 1}$

(viii) $y = \ln(\tan 2x)$

(ix) $y = \sin(\ln 3x)$

(x) $y = \tan(e^{x^2})$

(xi) $y = \sqrt{\cos 2x}$

(xii) $y = \sin^3(2e^{x-1})$

(xiii) $y = \cos^4\left[\ln(\sin e^x)\right]$

(xiv) $y = e^{\sin^2(\ln x)}$

2 (i) Differentiate $y = x^2 \sin x$.

(ii) Hence find the equation of the tangent to the curve at $x = \pi$.

3 Given that $f(x) = 2\sin^2 3x$, find the exact value of $f'\left(\dfrac{\pi}{18}\right)$.

4 Consider the equation of a curve $y = \dfrac{e^{2x}}{\cos x}$ for $x > 0$.

(i) Find $\dfrac{dy}{dx}$.

(ii) Hence find the x co-ordinate of the stationary point of the curve for $0 \leqslant x \leqslant \pi$.

5 The diagram shows the curve $y = 3\sin^2 x \cos^3 x$ for $0 \leqslant x \leqslant \dfrac{\pi}{2}$ and its maximum point N.

Find the x co-ordinate of N.

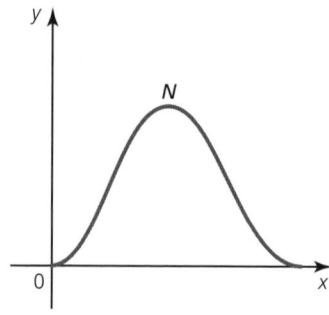

6 The curve $y = \dfrac{\cos 2x}{e^{2x}}$ has two stationary points for $0 \leqslant x \leqslant \pi$.

Find the x co-ordinates of these stationary points.

7 Consider the curve defined by $y = \ln(\cos 2x) + 2x$ for $0 \leq x \leq 2\pi$.

   (i) For what values of x is the function undefined for $0 \leq x \leq 2\pi$?

   (ii) Find the x co-ordinates of all the stationary points on the curve.

   (iii) Determine the nature of each of these stationary points.

Implicit differentiation

EXERCISE 4.4

1 Differentiate the following with respect to x.

(i) $2y^3$

(ii) $3x^2 - 5y^4 - 8$

(iii) $\sin 2x + \cos 2y$

(iv) e^{3y}

(v) $4x^2 y$

(vi) $\ln(xy)$

(vii) $\tan xy^2 - e^y$

(viii) $(e^{\sin y})^x$

2 Differentiate the following with respect to x and find an expression for $\dfrac{dy}{dx}$ in terms of x and y.

(i) $y^2 - 2x^3 = 5$

(ii) $x^2 y = \sin y$

(iii) $2y^3 + x = 4xy$

(iv) $e^{xy} - 2x = 12$

3 Find the equation of the normal to the curve $\ln(2xy) + y^2 = 1$ at the point $\left(\frac{1}{2}, 1\right)$.

4 The diagram shows the graph of the ellipse $4x^2 + y^2 = 8$.

Find the co-ordinates of the points on the ellipse where the gradient is 2.

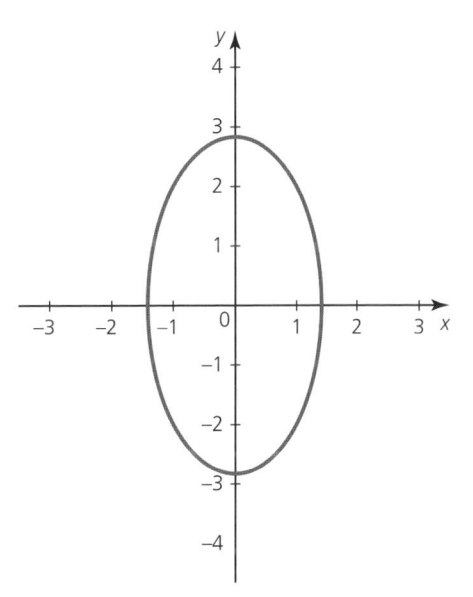

5 Find the co-ordinates of the point(s) on the curve $x^2 + 6y^2 - 2x + 8y = 39$ where the tangent to the curve is parallel to the x axis.

6 A curve is defined by $8 + x^2 = 2xy + y^2$.

(i) Find the co-ordinates of the stationary points on the curve.

(ii) Show that slope of the curve is never parallel to the y axis.

7 Find the co-ordinates of the points on the curve $x^2 + y^2 - 2x + 4y - 4 = 0$ where the tangent to the curve is parallel to the y axis.

8 The equation of a curve is $5x^2 - 2xy + 3y^2 - 70 = 0$.

(i) Show that $\dfrac{dy}{dx} = \dfrac{y - 5x}{3y - x}$.

(ii) Find the co-ordinates of each of the points on the curve where the tangent is parallel to the *x* axis.

(iii) Find the co-ordinates of each of the points on the curve where the tangent is parallel to the *y* axis.

9 The diagram shows the curve defined implicitly by
$y^2 + y = x^3 + 2x$.

(i) Find the co-ordinates of the points of intersection of the curve and the line $x = 2$.

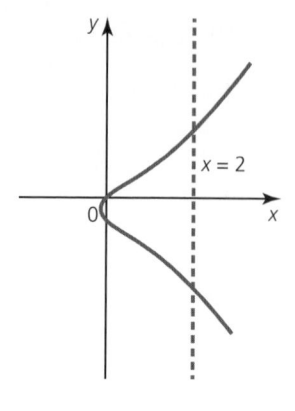

(ii) Find $\dfrac{dy}{dx}$ in terms of x and y and the gradient of the curve at these two points.

OCR MEI Structured Mathematics C3 4753/1 May 2005 Q7

10 Find the equation of the normal to the curve $x^3 + 4x^2y + y^3 = 6$ at the point $(1, 1)$, giving your answer in the form $ax + by + c = 0$, where a, b and c are integers.

11 The equation of a curve is $x^3 + y^3 = 6xy$.

   (i) Find $\dfrac{dy}{dx}$ in terms of x and y.

   (ii) Show that the point $\left(2^{\frac{4}{3}}, 2^{\frac{5}{3}}\right)$ lies on the curve and that $\dfrac{dy}{dx} = 0$ at this point.

(iii) The point (*a*, *a*), where *a* > 0, lies on the curve. Find the value of *a* and the gradient of the curve at this point.

Parametric differentiation

EXERCISE 4.5

1 Find $\dfrac{dy}{dx}$ in terms of t or θ for these curves defined parametrically.

(i) $x = 2t$
$y = t^2$

(ii) $x = 2\cos\theta$
$y = 3\sin\theta$

(iii) $x = \theta + \sin\theta$
$y = \theta - \cos 2\theta$

(iv) $x = 2t^3$
$y = (2t+1)^2$

(v) $x = 3e^t$
$y = t - e^{2t}$

(vi) $x = 2\sec\theta$
$y = 5\tan\theta$

2 A curve is defined parametrically by $x = \dfrac{t}{1-t}$, $y = \dfrac{t^2}{1+t}$.

(i) Find $\dfrac{dy}{dx}$.

(ii) Hence find the co-ordinates of the stationary points on the curve.

3 The parametric equations of a curve are $x = \ln(2-3t)$, $y = \dfrac{6}{t}$ for $t < 0$.

  (i) Show that $\dfrac{dy}{dx} = \dfrac{4-6t}{t^2}$.

  (ii) Find the exact co-ordinates of the only point on the curve where the gradient is equal to 4.

  (iii) Show that the cartesian equation of the curve is given by $y = \dfrac{18}{2-e^x}$.

4 A curve has parametric equations $x = e^{2t}$, $y = \dfrac{2t}{1+t}$.

Find the gradient of the curve at the point where $t = 0$.

OCR MEI Applications of Advanced Mathematics C4 4754/A January 2010 Q3

5 The graph shows the curve given by the parametric equations

$x = 2\sin\theta + \cos\theta,$
$y = \sin\theta + 2\cos\theta$.

(i) Find the equation of the tangent to the curve at the point where $\theta = \dfrac{\pi}{2}$.

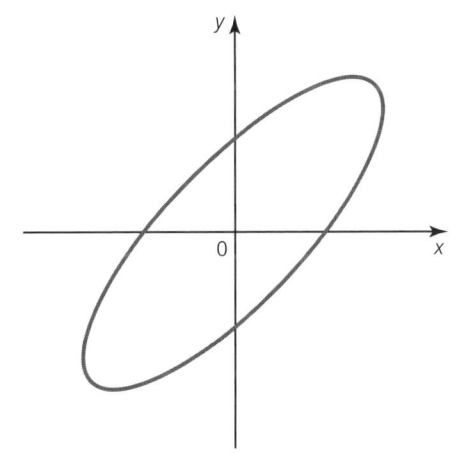

(ii) (a) Show that for any point of the curve $x^2 + y^2 = 5 + 4\sin 2\theta$.

(b) Hence find the greatest and least distance of a point on the curve from the origin.

6 This logo is defined by the parametric equations

$x = 3\sin t$,

$y = 3\cos 3t$

where t is time in seconds.

A point on the logo is illuminated at every moment of time t.

(i) Give the co-ordinates of the point when $t = 2$.

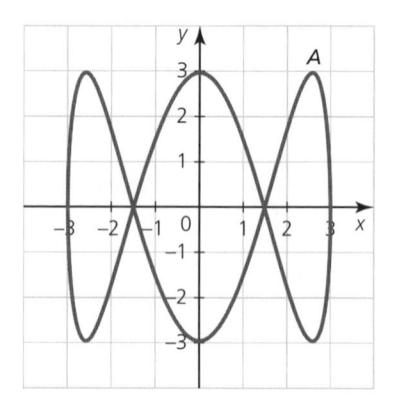

(ii) Label the point $M(3\sin 1, 3\cos 3)$ on the diagram.

(iii) How long does it take for the point to trace out the entire logo?

(iv) Find the slope of the curve when $t = 1$.

(v) Find the co-ordinates of the maximum point A shown on the curve.

(vi) Speed is defined as $\left|\dfrac{dy}{dx}\right|$. Find when the speed is a minimum the second time.

7 The parametric equations of a curve are $x = 2t - \ln t$, $y = t^2 - \ln t^2$ for $t \geq 0$.

  (i) Find $\dfrac{dy}{dx}$ and hence find the exact co-ordinates of the stationary point on the curve.

  (ii) Find the co-ordinates of the point where $\dfrac{dy}{dx} = 2$.

8 The parametric equations of a curve are $x = 2\theta + \sin 2\theta$, $y = 4\sin\theta$ and part of its graph is shown.

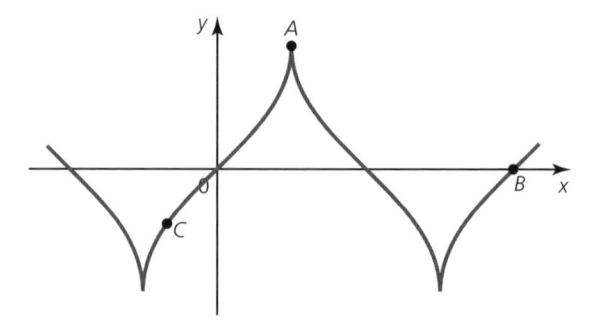

(i) Find the value of θ at A and the value of θ at B.

(ii) Show that $\dfrac{dy}{dx} = \sec\theta$.

(iii) At the point C on the curve, the gradient is 2.

Find the co-ordinates of C, giving your answer in an exact form.

OCR Advanced Mathematics Core Mathematics 4 4724/01 May 2008 Q9

9 In a theme park ride, a capsule C moves in a vertical plane (see diagram).

With respect to the axes shown, the path of C is modelled by the parametric equations

$x = 10\cos\theta + 5\cos 2\theta, \quad y = 10\sin\theta + 5\sin 2\theta,$

where x and y are in metres.

(i) Show that $\dfrac{dy}{dx} = -\dfrac{\cos\theta + \cos 2\theta}{\sin\theta + \sin 2\theta}$.

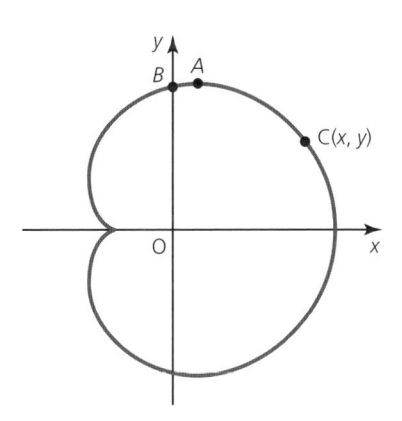

(ii) Hence find the exact co-ordinates of the highest point A on the path of C.

(iii) Express $x^2 + y^2$ in terms of θ.

Hence show that $x^2 + y^2 = 125 + 100\cos\theta$.

(iv) Using this result, or otherwise, find the greatest and least distances of C from O.

Stretch and challenge

1 The path traced out by a point on the circumference of a circle of radius r as the circle rolls along a straight line is called a *cycloid*.

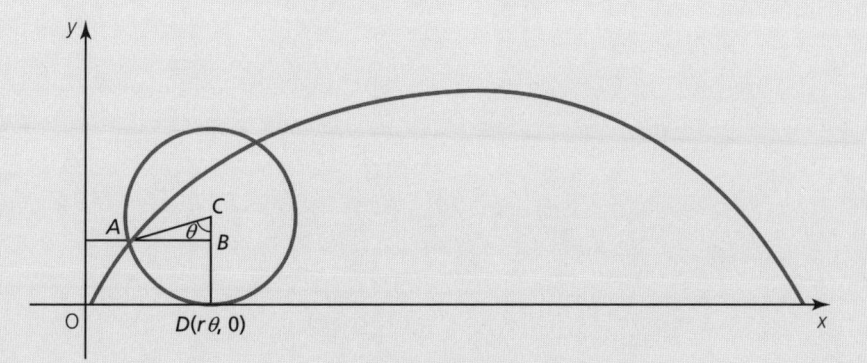

A is a point on the circumference that starts at the origin O.

The diagram shows the position of the circle after rotating through θ radians.

(i) Show that the parametric equations of a cycloid are given by:

$x = r(\theta - \sin\theta), \quad y = r(1 - \cos\theta) \quad 0 \leq \theta \leq 2\pi.$

(ii) Show that $\dfrac{dy}{dx} = \cot\left(\dfrac{\theta}{2}\right)$.

(iii) Find the equation of the tangent to the curve at the point where $y = \tfrac{1}{2}r$.

(iv) The speed of the point is given by $s = \sqrt{\left(\dfrac{dx}{d\theta}\right)^2 + \left(\dfrac{d}{d\theta}\right)^2}$.

Find and simplify an expression for the speed of the point.

(v) Given that acceleration, a, is given by $a = \dfrac{ds}{d\theta}$, find and simplify an expression for the acceleration of the point.

2 The parametric equations of a curve are $x = t^3$, $y = t^2$.

(i) Show that the equation of the tangent at the point P where $t = p$ is
$3py - 2x = p^3$.

(ii) Given that this tangent passes through the point $(-10, 7)$, find the co-ordinates of each of the three possible positions of P.

3 A ladder is being manoeuvred around a corner in a house.

One hallway is 2 m wide, the other hallway is 3 m wide.

(i) Find an expression for the length of the ladder in terms of θ.

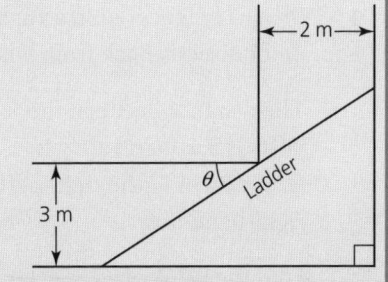

(ii) Hence find the maximum length the ladder can be so it will fit around the corner.

4 When a try is scored in a rugby game the kicker, K, must attempt a conversion from a point directly back from where the try was scored.

The distance between the posts is 5.6 m and a try is scored 30 m to the right of the right-hand post.

Find the value of x (how far back the kicker should go) so that the angle θ between the posts is a maximum.

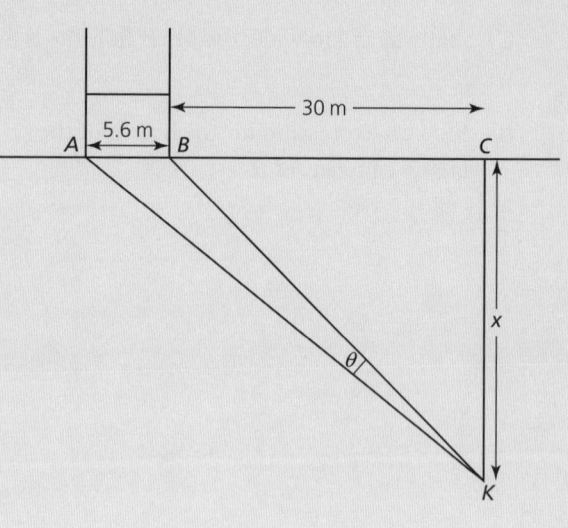

5 A three-dimensional solid has a surface made of two types of triangle, Type A and Type B.

The area of a Type A triangle is $A_A = \sqrt{2k^2 - 2k + 1}$.

The area of a Type B triangle is $A_B = \dfrac{\sqrt{3}}{2}(k^2 - k + 1)$.

Find the value of k which minimises the total surface area, and show that this value gives a minimum.

NZQA Scholarship Calculus 2011 Q1b

6 The nautilus is a marine creature that lives around coral reefs.

The mathematical model of a nautilus shell is an equiangular spiral.

Equiangular spirals have equations of the form $r = Ae^{k\theta}$, where k is a constant.

At every point P, the tangent to the curve makes the same angle, α, with the line OP from the point P to the origin (or pole), O.

The size of the angle α depends upon the number k in this mathematical model where $r = Ae^{k\theta}$.

(i) Using the parametric equations for the cartesian co-ordinates (x, y) of the point P in terms of θ, find $\dfrac{dy}{dx}$.

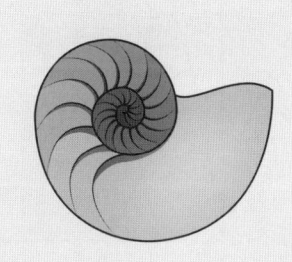

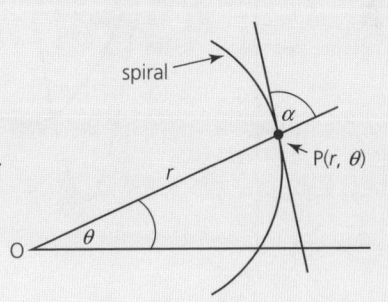

(ii) Hence, or otherwise, find the value of α in terms of k for this model.

NZQA Scholarship Calculus 2004 Q5

7 The graphs of the two functions $f(x) = 3\sin 2x$ and $g(x) = 2\cos x$ are shown.

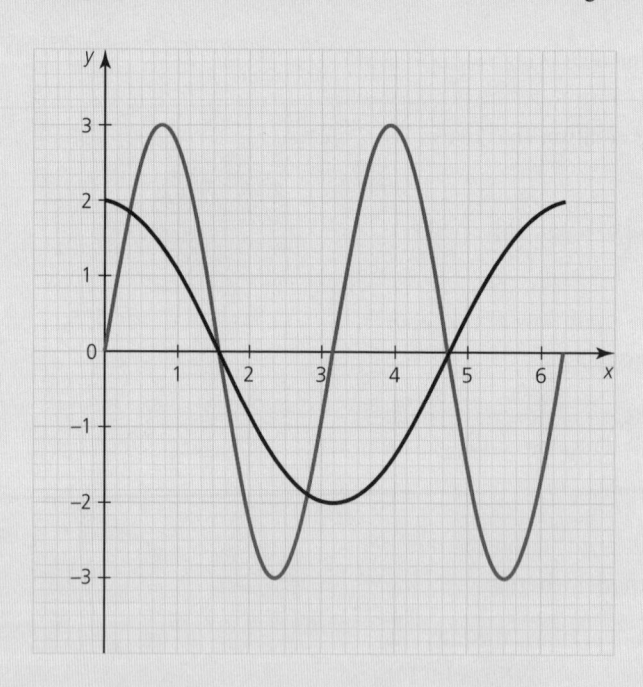

(i) The function h is defined by

h(x) = $k + 2\cos x$ for $0 \leq x \leq 2\pi$, where k is a positive constant.

Find the x co-ordinates of all the points at which the graphs of f(x) and h(x) have the same gradient.

(ii) The graph of p(x) = $-a(x-b)^2 - c$ for $0 \leq x \leq 2\pi$, where $a > 0$, $b > 0$, and $c > 0$, intersects each of the graphs of f(x) and g(x) at exactly one point only.

The x co-ordinate of the point of intersection with f(x) is 2.575 and the vertex of p(x) lies on g(x).

Find the values of a, b and c.

(iii) Find the range of values of k for which f(x) and h(x) intersect at exactly two points.

Note: $k > 0$ and both f(x) and h(x) have domain $0 \leq x \leq 2\pi$.

Exam focus

1 The diagram shows the curve $y = \sqrt{3+x^2}\,e^{-\frac{x}{4}}$.

Find the x co-ordinates of the stationary points on the curve. [4]

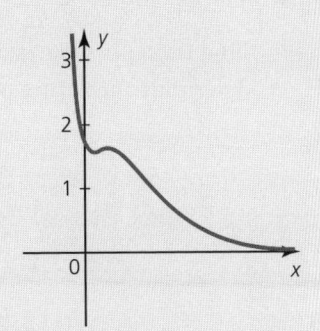

2 The curve shown is $y = \dfrac{(\ln x)^2}{x}$.

Find $\dfrac{dy}{dx}$ and hence find the exact co-ordinates of the maximum point M. [4]

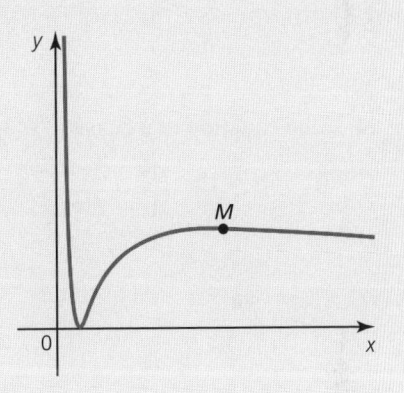

3 The equation of a curve is $y = \sin 2x + x$.

Find the co-ordinates of the stationary points on the curve for $0 \leq x \leq \pi$, and determine the nature of these stationary points. [5]

4 The equation of a curve is $x^2 + y^2 - xy - 48 = 0$.

  (i) Show that $\dfrac{dy}{dx} = \dfrac{2x-y}{x-2y}$ [4]

  (ii) Find the co-ordinates of the points on the curve where the tangent is parallel to the x axis. [3]

(iii) Find the co-ordinates of the points on the curve where the tangent is parallel to the y axis. [3]

5 The equation of a curve is $2x^2 + xy + y^2 = 14$.

Show that there are two stationary points on the curve and find their co-ordinates. [6]

6 The parametric equations of a curve are $x = \dfrac{2t}{3t+4}$, $y = 3\ln(3t+4)$.

(i) Express $\dfrac{dy}{dx}$ in terms of t, simplifying your answer. [4]

(ii) Find the gradient of the curve at the point for which $x = 2$. [2]

5 Integration

Integrals involving the exponential and natural logarithm functions

EXERCISE 5.1

1 Find the following integrals.

(i) $\int e^{2x} \, dx$

(ii) $\int e^{1-3x} \, dx$

(iii) $\int \dfrac{1}{2x} \, dx$

(iv) $\int \dfrac{2}{x} \, dx$

(v) $\int \dfrac{4}{e^{2x+3}} \, dx$

(vi) $\int \dfrac{3}{2x+1} \, dx$

(vii) $\int \dfrac{e^{4x}+1}{e^{x}} \, dx$

(viii) $\int 9 e^{-\frac{1}{3}x} \, dx$

(ix) $\int (e^{x} - 1)^2 \, dx$

(x) $\int \dfrac{1}{3x} + \dfrac{1}{1-2x} \, dx$

2 Find the following definite integrals, giving your answer exactly.

(i) $\int_1^2 3 e^{4x} \, dx$

(ii) $\int_2^4 \dfrac{2}{x+1} \, dx$

3 Show that $\int_1^{\infty} \dfrac{e^x + 2}{e^{2x}} \, dx = \dfrac{e+1}{e^2}$.

4 Show that $\int_{-1}^{1} \dfrac{9}{1-3x} dx = \ln 8$.

5 Find $\int \dfrac{x-1}{x^2-1} dx$.

6 Find $\int \dfrac{3x^2-7x}{3x+2} dx$.

7 The area between the curve $y = \dfrac{3}{3x-1}$ and the x axis between $x = k$ and $x = 1$ is exactly 1. Find the exact value of k.

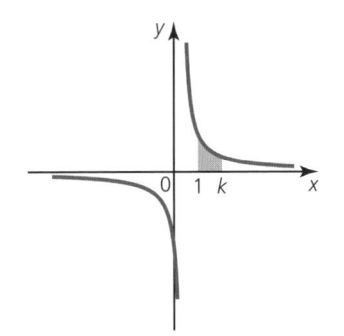

8 Given that

$$\int_0^{\ln 4} \left(ke^{3x} + (k-2)e^{-\frac{1}{2}x} \right) dx = 185$$

find the value of the constant k.

Integrals involving trigonometrical functions

EXERCISE 5.2

1 Find the following integrals.

(i) $\int \sin 4x \, dx$

(ii) $\int \cos(3x-1) \, dx$

(iii) $\int \sec^2 2x \, dx$

(iv) $\int 2\sin\tfrac{1}{3}x \, dx$

(v) $\int \left(4\cos\tfrac{1}{2}x + 1\right) dx$

(vi) $\int (\cos 2x - \sec^2 3x + 4) \, dx$

2 Find the exact values of the definite integrals.

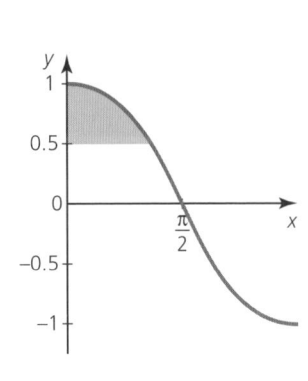

(i) $\int_0^{\frac{\pi}{4}} \sin 2x \, dx$

(ii) $\int_0^{\frac{\pi}{16}} (1 - \cos 4x) \, dx$

(iii) $\int_{\frac{\pi}{3}}^{\frac{\pi}{2}} \sec^2 \tfrac{1}{2}x \, dx$

(iv) $\int_0^{\frac{\pi}{4}} \tan^2 x \, dx$

3 Find the area enclosed between the curve $y = \cos x$, the line $y = \tfrac{1}{2}$ and the y axis for $0 \leqslant x \leqslant \tfrac{\pi}{2}$.

4 Find the area between the curve $y = \sin x$ and the line $y = \frac{2}{\pi}x$ for $x \geqslant 0$.

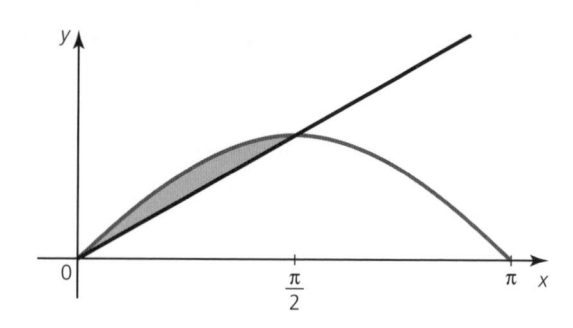

5 Find the integrals by using a trigonometric identity.

(i) $\int \sin^2 x \, dx$

(ii) $\int \cos^2 2x \, dx$

(iii) $\int 4\sin x \cos x \, dx$

(iv) $\int \cos^2 2x - \sin^2 2x \, dx$

(v) $\int \sin^4 x \, dx$

(vi) $\int (\sin x - \cos x)^2 \, dx$

(vii) $\int (\sin 4x \cos x - \cos 4x \sin x) \, dx$

(viii) $\int \cos^4 x \, dx$

6 Show that $\int_0^{\frac{\pi}{4}} \sqrt{1+\cos 4x}\, dx = \frac{\sqrt{2}}{2}$.

7 The diagram shows the curve $y = 2\sin x$ for $0 \leqslant x \leqslant \pi$ and the line $y = 1$.

   (i) Find the area enclosed by the curve and the line.

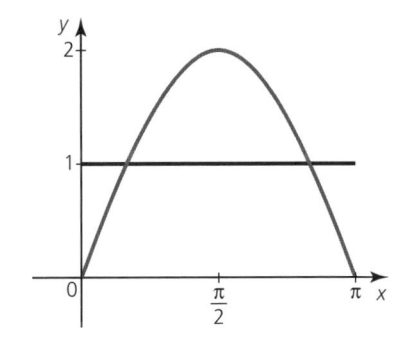

   (ii) The region enclosed by the curve and the line is rotated 360° around the x axis. Find the volume of the solid generated.

8 (i) Show that $\cos 3x \equiv 4\cos^3 x - 3\cos x$.

(ii) Hence find the exact value of $\int_{\frac{\pi}{3}}^{\frac{\pi}{2}} \cos^3 x \, dx$.

9 (i) Given that $5\cos x - 3\sin x = A(\cos x + \sin x) + B(\cos x - \sin x)$ for all values of x, find the values of the constants A and B.

(ii) Hence find the exact value of $\int_0^{\frac{\pi}{2}} \dfrac{5\cos x - 3\sin x}{\cos x + \sin x}\,dx$.

10 By expressing $\cos 2x$ in terms of $\cos x$, find the exact value of $\int_{\frac{1}{6}}^{\frac{1}{3}} \dfrac{\cos 2x}{\cos^2 x}\,dx$.

Numerical integration

EXERCISE 5.3

1 The values of x and $f(x)$ are given in the table below for $0 \leqslant x \leqslant 2$.

x	0	0.5	1	1.5	2
$f(x)$	2.1	0.8	1.5	1.9	3.8

Use the trapezium rule with 4 strips to find an estimate for $\int_0^2 f(x)\,dx$.

2 The function $f(x)$ is defined by $f(x) = \dfrac{1}{x^3+1}$.

(i) Complete the table of values for $f(x)$.

x	0	0.25	0.5	0.75	1
$f(x)$	1				0.5

(ii) Use the trapezium rule with 4 strips to estimate the area between the curve and the x axis between $x = 0$ and $x = 1$.

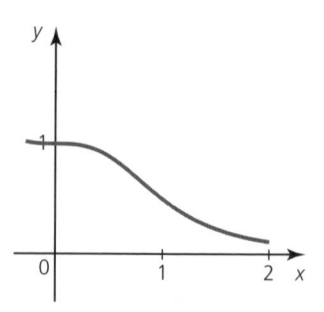

(iii) Using the graph, determine whether the answer to part **(ii)** is an overestimate or underestimate of the true area under the curve.

3 The diagram shows the cross-section of a certain section of river.
The depth of the river is measured as shown.
The measurements are made 0.5 m apart.

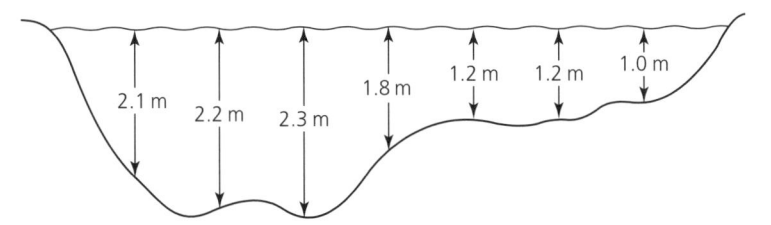

(i) Use the trapezium rule to find an estimate of the area of the cross-section.

(ii) If the river is flowing at a constant 3 km/h, find the volume of water in m³ passing this point of the river every minute.

4 The diagram shows the curve $y = \dfrac{e^x}{x^2}$ for $0 < x < 5$.

Use the trapezium rule with strip widths 0.5 to find an estimate for $\displaystyle\int_1^3 \dfrac{e^x}{x^2}\,dx$ correct to 2 decimal places.

Is your answer an underestimate or overestimate of the true area?

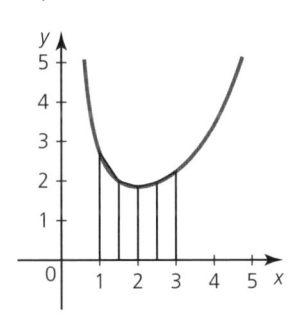

x	1	1.5	2	2.5	3
y	2.718				

5 Use the trapezium rule with two intervals to estimate $\displaystyle\int_0^2 \dfrac{2}{8+3e^x}\,dx$.

6 The diagram shows part of the curve $y = \sqrt{1+x^3}$.

(i) Use the trapezium rule with 4 strips to estimate $\int_0^2 \sqrt{1+x^3}\,dx$, giving your answer correct to 3 significant figures.

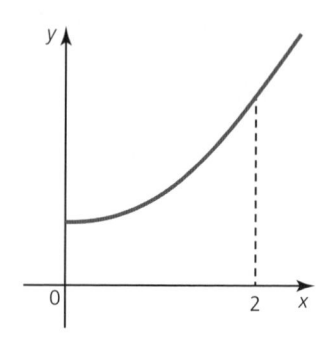

(ii) Chris and Dave each estimate the value of this integral using the trapezium rule with 8 strips.

Chris gets a result of 3.25, and Dave gets 3.30. One of these results is correct.

Without performing the calculation, state with a reason which is correct.

Stretch and challenge

1. The length of a curve between $x = b$ and $x = a$ is given by the formula
$$L = \int_a^b \sqrt{1+[f'(x)]^2}\, dx.$$

   Show that the length of the curve $y = \dfrac{x^3}{6} + \dfrac{1}{2x}$ between $x = 2$ and $x = 1$ is $\dfrac{17}{12}$.

2. Show that $\int_0^{\frac{\pi}{4}} \sin^2 x \cos^2 x\, dx = \dfrac{\pi}{32}$.

3 A curve in polar form is given in terms of its distance from the origin (r) and the angle made with the positive x axis (θ).

A point with $r = 2$ and $\theta = \dfrac{\pi}{6}$ is given in polar form by $\left(2, \dfrac{\pi}{6}\right)$.

Some polar curves are shown below.

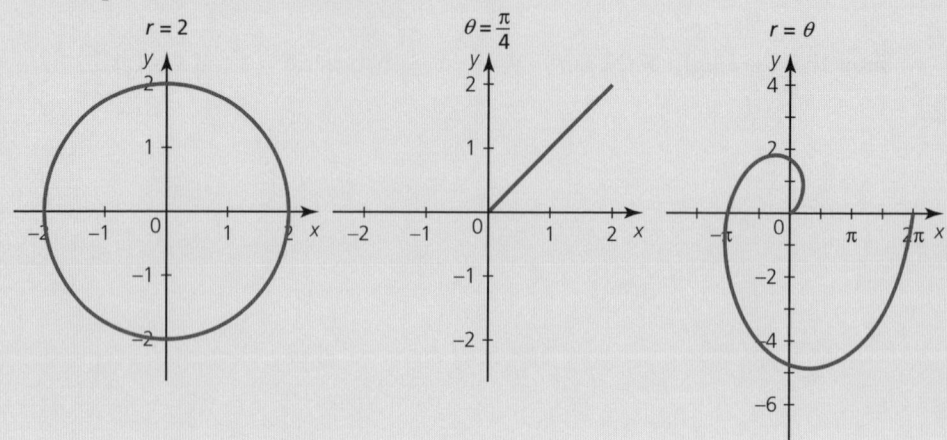

(i) The diagram shows the polar curves $r = k\theta$, $\theta = m$ and $r = t$ (where k, m and t are constants).

If point P has polar co-ordinates $\left(\dfrac{3}{2}, \dfrac{3}{4}\right)$, determine the exact values of k, m and t.

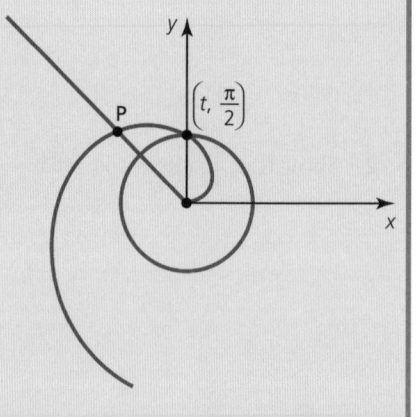

(ii) If $r = p(\theta)$, the length of a polar curve is given by

$$L = \int_0^2 \sqrt{[r(\theta)]^2 + [r'(\theta)]^2}\, d\theta.$$

Find the length of the cardioid $r = 1 + \cos\theta$.

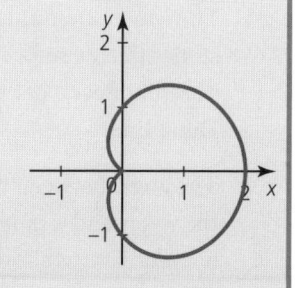

(iii) The parametric equations of a cycloid are given by:

$$x = r(\theta - \sin\theta), \quad y = r(1 - \cos\theta) \quad 0 \leq \theta \leq 2.$$

Find the length of the cycloid which is given by:

$$L = \int_0^2 \sqrt{\left(\frac{dx}{d\theta}\right)^2 + \left(\frac{dy}{d\theta}\right)^2}\, d\theta.$$

4 The Moeraki boulders are natural stone spheres sunk into the sand of Moeraki Beach between Oamaru and Dunedin.

The angle between the surface of the water and a tangent plane to the boulder is ϕ, as shown.

Find the proportion of the volume of the boulder which is below water level.

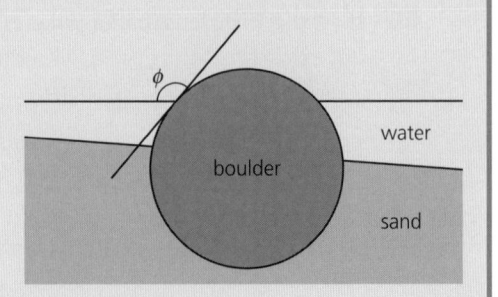

NZQA Scholarship Calculus 2011 Q3c

Exam focus

1 (i) Show that $\int_0^2 \frac{4}{2x+1} dx = \ln 25$. [3]

(ii) Find the value of k such that $\int_k^2 \frac{4}{2x+1} dx = 0$. [4]

2 (i) Show that $(2\cos x + \sin x)^2$ can be written in the form $a\sin 2x + b\cos 2x + c$, stating the values of a, b and c. [4]

(ii) Hence find the exact value of $\int_0^{\frac{\pi}{2}} (2\cos x + \sin x)^2 dx$. [4]

3 Show that $\tan^2 x + \sin^2 x = \sec^2 x - \frac{1}{2}\cos 2x - \frac{1}{2}$ and hence find the exact value of $\int_0^{\frac{\pi}{6}} (\tan^2 x + \sin^2 x)\,dx$. [6]

4 The diagram shows a part of the curve $y = \sqrt{4 - \cos x}$ for $0 \leqslant x \leqslant \pi$.

   (i) Use the trapezium rule with two intervals to estimate the value of $\int_0^{\pi} \sqrt{4 - \cos x}\,dx$ correct to 3 significant figures. [3]

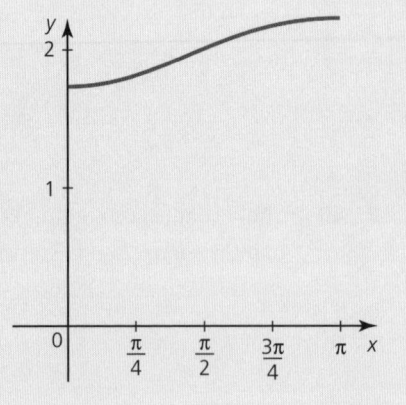

   (ii) Explain, with reference to the diagram, why the trapezium rule may be expected to give a good approximation to the true value of the integral in this case. [1]

5 (i) Differentiate $e^x(\sin 2x - 2\cos 2x)$, simplifying your answer. [3]

(ii) Hence find the exact value of $\int_0^{\frac{1}{4}} e^x \sin 2x \, dx$. [3]

6 (i) Express $\cos\theta + \sqrt{3}\sin\theta$ in the form $R\cos(\theta - \alpha)$ where $R > 0$ and α is acute. Express α in terms of π. [3]

(ii) Hence show that $\int_0^{\frac{1}{3}} \frac{1}{(\cos\theta + \sqrt{3}\sin\theta)^2} \, d\theta = \frac{\sqrt{3}}{4}$. [4]

6 Numerical solution of equations

Interval estimation – change-of-sign methods

EXERCISE 6.1

1 The diagram shows the graphs of $y = e^{x-3}$ and $y = x^3$.

To find where the two curves intersect, we solve the equation $e^{x-3} = x^3$.

(i) Rearrange the equation so it is in the form $f(x) = 0$.

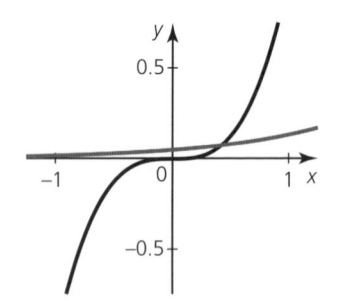

(ii) Show that the root of $f(x) = 0$ lies between 0 and 1.

(iii) Show that the root of $f(x) = 0$ lies between 0 and 0.5.

(iv) Find the two values of x, correct to 1 decimal place, between which the root lies.

2 The graph of the curve $f(x) = x^3 - 2x^2 - 4x + 6$ is shown.

Using the change of sign idea, find the integer bounds for each of the three roots of the equation $f(x) = 0$.

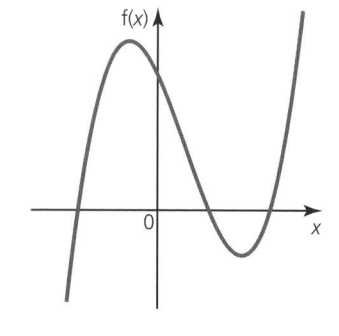

3 For the curve $g(x) = \dfrac{x+4}{x-1}$, $g(0) = -4$ and $g(2) = 6$.

As there is a change in sign between $x = 0$ and $x = 2$, there must be a root of the equation $g(x) = 0$ between $x = 0$ and $x = 2$.

State, with a reason, why this statement is false in this case.

4 (i) By drawing a sketch, show that the equation $4 - x = \ln x$ has only one root.

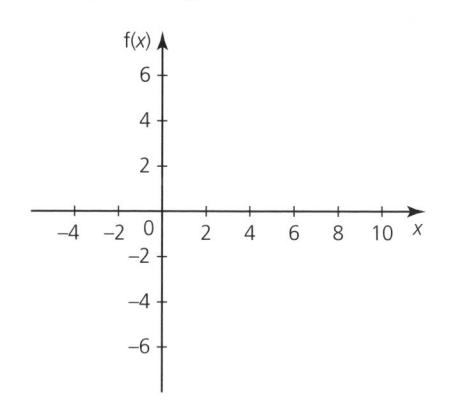

(ii) Rearrange the equation $4 - x = \ln x$ so it is in the form $f(x) = 0$.

(iii) Find the integers a and b such that $f(a)>0$ and $f(b)<0$ or $f(a)<0$ and $f(b)>0$.

Hence state the integer bounds between which the root of $f(x)=0$ lies.

(iv) Verify by calculation that this root lies between 2.9 and 3.0.

5 (i) Show that the equation $x^2 = 3^x$ has a root α in the interval $-0.7 < \alpha < -0.6$.

(ii) Sketch the graphs of $y = x^2$ and $y = 3^x$ to verify there is just one root of the equation $x^2 = 3^x$.

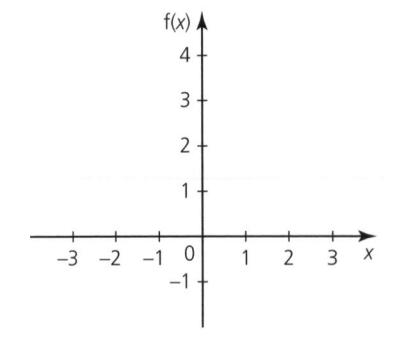

Fixed-point iteration

EXERCISE 6.2

1 The equation $x^3 - 2x + 3 = 0$ has one root.

  (i) Sketch the graphs of $y = x^3$ and $y = 2x - 3$ on the grid below.

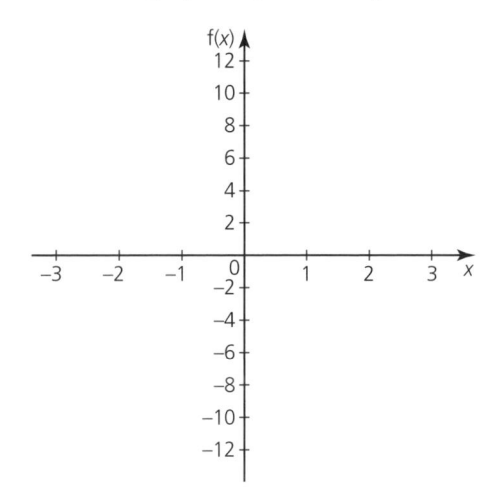

  (ii) Use your graph to find the integers between which the root lies.

  (iii) Use the iterative formula $x_{n+1} = \sqrt[3]{2x_n - 3}$ to find the root correct to 4 decimal places.

  (iv) Suggest two other arrangements of the form $x_{n+1} = f(x_n)$ that could be used to find the root.

2 The equation $4 - x = \ln x$ has a root α where $2.9 < \alpha < 3.0$.

   Use the iterative formula $x_{n+1} = 4 - \ln x_n$ with initial value $x_0 = 2.9$ to find the value of α correct to 3 decimal places.

   Give the result of each iteration to 4 decimal places.

3 **(i)** Show that the equation $x^3 - x^2 = 15$ has a root between $x = 2$ and $x = 3$.

   (ii) Use the iterative formula $x_{n+1} = \sqrt[3]{15 + x_n^2}$ with $x_0 = 2.5$ to find the root correct to 3 decimal places.

4 **(i)** Show that if the iterative formula $x_{n+1} = \sqrt{\dfrac{3 - x_n}{x_n}}$ converges to the value α, then α will be a root of the equation $x^3 + x - 3 = 0$.

(ii) Use the iterative formula with $x_0 = 1.5$ to find the value of α correct to 2 decimal places.

5 The diagram shows a shaded segment of a circle centre O radius r.

   (i) Show that the area, S, of the segment is given by
   $S = \frac{1}{2}r^2(\theta - \sin\theta)$.

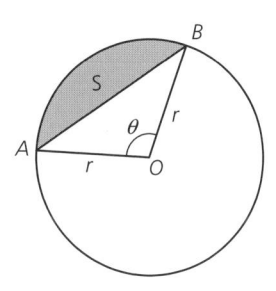

   (ii) The cord AB divides the area of the circle in the ratio $1:5$.

   Show that θ satisfies $\theta = \frac{1}{3} + \sin\theta$.

   (iii) Use the iterative formula
   $$\theta_{n+1} = \frac{1}{3} + \sin\theta_n$$
   with $\theta_1 = 1$ to find θ correct to 2 decimal places.

6 The equation $x^2 = 3^x$ has a root α in the interval $-0.7 < \alpha < -0.6$.

Show that each of these three possible arrangements $x_{n+1} = f(x_n)$ fail to converge to the root.

(i) $x_{n+1} = \sqrt{3^{x_n}}$

(ii) $x_{n+1} = \log_3(x_n^2)$

(iii) $x_{n+1} = \dfrac{3^{x_n}}{x_n}$

7 (i) Given that $\int_0^a (6e^{2x} + x)\,dx = 42$, show that $a = \tfrac{1}{2}\ln\left(15 - \tfrac{1}{6}a^2\right)$.

(ii) Use an iterative formula, based on the equation in part (i), to find the value of a correct to 3 decimal places.

Use a starting value of 1 and show the result of each iteration to 4 decimal places.

OCR Core Mathematics 3 4723/01 June 2007 Q6

8 The sequence defined by

$$x_1 = 3, \quad x_{n+1} = \sqrt[3]{31 - \tfrac{5}{2}x_n}$$

converges to the number α.

(i) Find the value of α correct to 3 decimal places, showing the result of each iteration.

(ii) Find an equation of the form $ax^3 + bx + c = 0$, where a, b and c are integers, which has α as a root.

OCR Core Mathematics 3 4723/01 January 2008 Q2

Stretch and challenge

1 The *secant method* is another way to find the root of an equation.

It requires two starting points, x_1 and x_2, but they need not be on opposite sides of the exact solution.

A straight line is drawn through the two points $(x_1, f(x_1))$ and $(x_2, f(x_2))$, and the next estimate is taken as the point at which this line cuts the x axis.

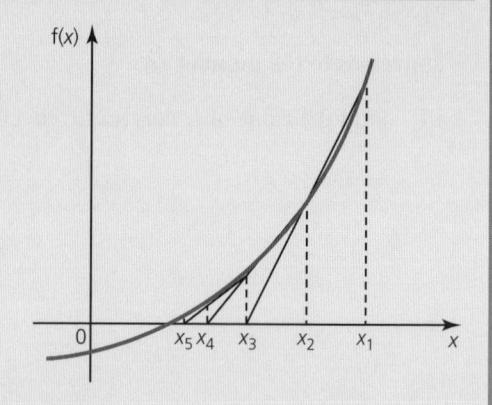

(i) Develop an equation for x_{n+1} in terms of x_n and x_{n-1}.

(ii) Use the secant method to find x_3 for the equation $f(x) = e^{2x} - 3$, given that $x_1 = 1$ and $x_2 = 0$.

Exam focus

1 The line $y = x$ intersects the curve $y = \sqrt{4 - \cos x}$ at the point M.

Use the iterative formula

$$x_{n+1} = \sqrt{4 - \cos x_n}$$

with $x_0 = 5$ to determine the x co-ordinate of M correct to 2 decimal places.

Give the result of each iteration to 4 decimal places. [3]

2 A curve is given by $y = e^{-\frac{1}{4}x}\sqrt{3 + x^2}$.

(i) The sequence of values given by the iterative formula

$$x_{n+1} = 2\ln(48 + 16x_n^2)$$

with initial value $x_0 = 14$ converges to a certain value α.

State an equation satisfied by α and hence show that α is the x co-ordinate of a point on the curve where $y = 0.25$. [3]

(ii) Use the iterative formula to calculate the value of α to 2 decimal places. Give the result of each iteration to 4 decimal places. [3]

3 The diagram shows the curve
$$y = x^4 - 4x^3 + 4x^2 + 2x - 7,$$
which crosses the x axis at the points $(\alpha, 0)$ and $(\beta, 0)$ where $\alpha < \beta$.

It is given that α is an integer.

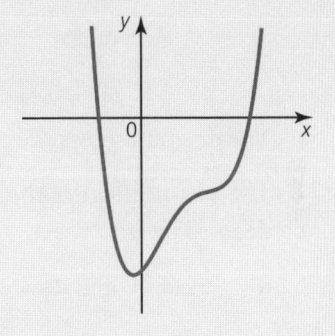

(i) Find the value of α. [2]

(ii) Show that β satisfies the equation $x = \sqrt[3]{5x^2 - 9x + 7}$. [3]

(iii) Use an iteration process based on the equation in part **(ii)** to find the value of β correct to 2 decimal places.

Show the result of each iteration to 4 decimal places. [3]

4 The diagram shows the curve $y = \dfrac{\cos 2x}{1-x}$.

The x co-ordinate of the maximum point M is denoted by α.

(i) Find $\dfrac{dy}{dx}$ and show that α satisfies the equation

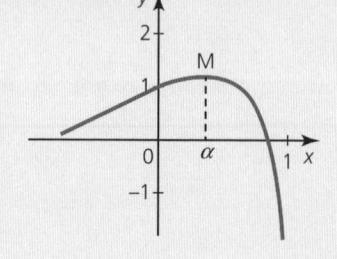

$$\tan 2x = \dfrac{1}{2-2x}.$$ [4]

(ii) Show by calculation that α lies between 0.3 and 0.4. [2]

(iii) Use the iterative formula $x_{n+1} = \frac{1}{2}\tan^{-1}\left(\frac{1}{2-2x_n}\right)$ to find the value of α correct to 3 decimal places.

Give the result of each iteration to 5 decimal places. [3]

5 The diagram shows a circle radius r centre O with the radius OB extended to meet the tangent to the circle at A at the point C. The shaded area is the same as the area of the sector OAB.

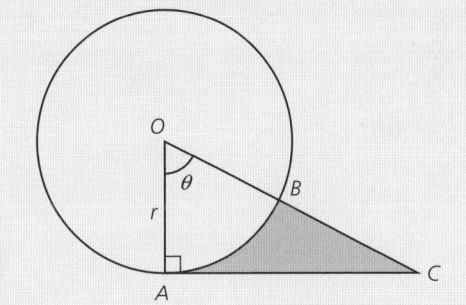

(i) Show that θ satisfies the equation
$2\theta = \tan\theta$. [2]

(ii) This equation has one root in the interval $0 < \theta < \frac{\pi}{2}$.

Use the iterative formula
$\theta_{n+1} = \tan^{-1}(2\theta_n)$

to find the root correct to 2 decimal places.

Give the result of each iteration correct to 4 decimal places. [3]

P3 7 Further algebra

The general binomial expansion

EXERCISE 7.1

1. Find the first 4 terms, in ascending powers of x, in the expansions below.

   In each case state the values of x for which the expansion is valid.

   (i) $(1+x)^{-2}$ **(ii)** $(1-2x)^{-1}$

   (iii) $(1+4x)^{\frac{1}{2}}$ **(iv)** $(1-9x)^{-\frac{1}{3}}$

2. Expand each of the following in ascending powers of x, as far as the term in x^2.

   Give the values of x for which each expansion is valid.

   (i) $(2+x)^{-4}$ **(ii)** $(9-3x)^{\frac{1}{2}}$

   (iii) $\dfrac{1}{\sqrt{4-x}}$ **(iv)** $\dfrac{3x}{(2+x)}$

3 The first three terms, in ascending powers of x, in the expansion of $(1 + ax)^b$ are $1 - 10x + 75x^2$.

  (i) Find the values of a and b.

  (ii) State the values of x for which the expansion is valid.

4 (i) Find the first three terms, in ascending powers of x, in the expansion of $\dfrac{1}{\sqrt{1-2x^2}}$.

State the set of values of x for which the expansion is valid.

(ii) Hence find the first six terms, in ascending powers of x, in the expansion of $\dfrac{1+x}{\sqrt{1-2x^2}}$.

5 In the expansion of $(1-3x)^n$, the coefficients of the x and x^2 terms are the same.

(i) Find the value of n.

(ii) When n has this value obtain the expansion up to and including the term in x^3, simplifying the coefficients.

6 When $(1 - 3x)(1 + ax)^{-2}$, where a is a constant ($a \neq 0$), is expanded in ascending powers of x, the coefficient of the term in x^2 is zero.

(i) Find the value of a.

(ii) When a has this value, find the term in x^3 in the expansion of $(1-3x)(1+ax)^{-2}$, simplifying the coefficient.

7 (i) Expand $(1-3x)^{-\frac{1}{3}}$ in ascending powers of x, up to and including the term in x^3.

(ii) Hence find the coefficient of x^3 in the expansion of $(1-3(x+x^3))^{-\frac{1}{3}}$.

8 (i) Expand $(1+ax)^{-4}$ in ascending powers of x, up to and including the term in x^2.

(ii) The coefficients of x and x^2 in the expansion of $(1+bx)(1+ax)^{-4}$ are 1 and -2 respectively.

Given that $a > 0$, find the values of a and b.

9 (i) Expand $(a + x)^{-2}$ in ascending powers of x up to and including the term in x^2.

(ii) When $(1-x)(a+x)^{-2}$ is expanded, the coefficient of x^2 is 0.

Find the value of a.

10 (i) Given that $y = \dfrac{1}{\sqrt{1-2x}-\sqrt{1-x}}$ where $x < \dfrac{1}{2}$ show that, assuming $x \neq 0$,

$y = -\dfrac{1}{x}(\sqrt{1-2x}+\sqrt{1-x})$.

(ii) Hence find the coefficient of the x term in the expansion of y.

Algebraic fractions review

EXERCISE 7.2

1 Simplify the following expressions.

(i) $\dfrac{3a}{4b} \times \dfrac{8b^2}{9a^2}$

(ii) $\dfrac{6c^2 d}{5e} \div \dfrac{18d^2}{25e^2}$

(iii) $\dfrac{f^2 - 16}{f^2 - 6f + 8}$

(iv) $\dfrac{3g+1}{4} \times \dfrac{g+1}{3g^2 + 4g + 1}$

(v) $\dfrac{h^3 - 4h}{4} \div \dfrac{h^2 - 4h + 4}{8}$

(vi) $\dfrac{j^4 - k^4}{j^2 + k^2} \times \dfrac{j+k}{j-k}$

2 Write each of these expressions as a single fraction in its simplest form.

(i) $\dfrac{2}{m} + \dfrac{3}{4m}$

(ii) $\dfrac{5n}{3} - \dfrac{n+1}{4}$

(iii) $\dfrac{p-2}{p} + \dfrac{p+2}{3}$

(iv) $\dfrac{2q}{q+1} - \dfrac{4}{q-1}$

(v) $\dfrac{r+5}{4r^2} + \dfrac{3}{5r} - \dfrac{2}{r^3}$

(vi) $\dfrac{s^2 + 4}{s^2 - 9} - \dfrac{s}{s-3}$

Partial fractions

EXERCISE 7.3

1 Express the following quotients as the sum of partial fractions.

(i) $\dfrac{5x+7}{(x+1)(x+2)}$

(ii) $\dfrac{2x+14}{(x-1)(x+3)}$

(iii) $\dfrac{3x-2}{x^2-4}$

(iv) $\dfrac{16-3x}{x^2+x-6}$

(v) $\dfrac{4}{x^2-x}$

(vi) $\dfrac{x-10}{2x^2-5x-3}$

(vii) $\dfrac{42-18x}{(x+1)(x-2)(x-4)}$

(viii) $\dfrac{5x^2+20x-32}{x^3-16x}$

(ix) $\dfrac{2x^3-4x^2-x-3}{x^2-2x-3}$

(Hint: Divide out first.)

(x) $\dfrac{3x^2+2x-20}{x^2-4}$

(Hint: Divide out first.)

2 Express the following fractions as the sum of partial fractions.

(i) $\dfrac{x^2+3x-1}{x^2(x-1)}$

(ii) $\dfrac{3x^2+2x-3}{(x-1)(x+1)^2}$

(iii) $\dfrac{1}{x(x-1)^2}$ (iv) $\dfrac{x-1}{(x+1)^3}$

3 Write the following fractions as the sum of partial fractions.

(i) $\dfrac{1}{(x+1)(x^2+1)}$ (ii) $\dfrac{2x^2+3x+1}{(x-1)(x^2+2)}$

(iii) $\dfrac{10}{(x-4)(x^2+4)}$ (iv) $\dfrac{-2x+4}{(x^2+1)(x-1)^2}$

Using partial fractions with the binomial expansion

EXERCISE 7.4

1 (i) If $\dfrac{2-11x}{(1+2x)(2-x)} = \dfrac{A}{1+2x} + \dfrac{B}{2-x}$, find the values of A and B.

(ii) Expand $\dfrac{1}{1+2x}$ as far as the term in x^2.
Give the range of values of x that the expansion is valid for.

(iii) Expand $\dfrac{1}{2-x}$ as far as the term in x^2.
Give the range of values of x that the expansion is valid for.

(iv) Hence find the first three terms, in ascending powers of x, in the expansion of $\dfrac{2-11x}{(1+2x)(2-x)}$, giving the values of x for which the expansion is valid.

2 Express the following in partial fractions and hence find the first three terms of the binomial expansion, stating the values of x for which this is valid.

(i) $\dfrac{8-x}{(x-2)(x+1)}$

(ii) $\dfrac{x^2+9x+2}{(1-2x)(1+x)^2}$

(iii) $\dfrac{2x^2-3x+3}{(1+x)(1+x^2)}$

3 (i) Express $\dfrac{4x+14}{(1-x)(2+x)(1+x)}$ in partial fractions.

(ii) Hence obtain the expansion of $\dfrac{4x+14}{(1-x)(2+x)(1+x)}$ in ascending powers of x, up to and including the term in x^2.

4 (i) Given that
$$\dfrac{3+2x^2}{(1+x)^2(1-4x)} \equiv \dfrac{A}{1+x} + \dfrac{B}{(1+x)^2} + \dfrac{C}{1-4x}$$
where A, B and C are constants, find B and C and show that $A = 0$.

(ii) Given that x is sufficiently small, find the first three terms in the binomial expansion of $\dfrac{3+2x^2}{(1+x)^2(1-4x)}$.

OCR MEI Structured Mathematics C4 4754(A) June 2006 Q2

Stretch and challenge

1 (i) Expand $\sqrt{4-3x}$ in ascending powers of x up to and including the term in x^2, simplifying the coefficients.

(ii) Find the value of the constant a such that the coefficient of the x^2 term in the expansion of $(1+ax)(4-3x)^{\frac{1}{2}}$ is $\frac{111}{64}$.

2 The coefficient of the x^2 term in the expansion of $\dfrac{(1+3x)^2}{(1+kx)^2}$ is 105. Find the value(s) of the constant k.

3 In the expansion of $(1 + x)^n$, where n is any rational number, for what values of n is the coefficient of the x^3 term less than the coefficient of the x^4 term?

4 (i) Given the binomial expansion $(1 + ax)^b = 1 - x + 2x^2 - \ldots$ find the values of a and b.

(ii) Hence state the set of values of x for which the expansion is valid.

Exam focus

1. When $(1 + ax)^{-3}$, where a is a positive constant, is expanded in ascending powers of x, the coefficients of the x and x^2 terms are the same.

   (i) Find the value of a. [4]

   (ii) When a has the value found in part (i), obtain the expansion up to and including the term in x^2, simplifying the coefficients. [3]

2. (i) Express $\dfrac{2-4x-3x^2}{(2-x)(2+x^2)}$ in partial fractions. [5]

   (ii) Hence obtain the expansion of $\dfrac{2-4x-3x^2}{(2-x)(2+x^2)}$ in ascending powers of x, up to and including the term in x^3. [5]

8 Further integration

Integration by substitution

EXERCISE 8.1

1 Use the given substitution to find the integrals.

(i) $\int (2x+1)\,dx \qquad u = 2x+1$

(ii) $\int 2x(x^2+1)^4\,dx \qquad u = x^2+1$

(iii) $\int x\sqrt{1-x^2}\,dx \qquad u = 1-x^2$

(iv) $\int \dfrac{x^2}{(3+2x^3)}\,dx \qquad u = 3+2x^3$

(v) $\int 2x\sqrt{x+4}\,dx \qquad u = x+4$

(vi) $\int \dfrac{2x}{(4-x)^5}\,dx \qquad u = 4-x$

(vii) $\int x^2\sqrt{x-1}\,dx \qquad u = x-1$

2 Find the following definite integrals using a suitable substitution.

(i) $\int_0^1 x^3 \sqrt{x^4+1}\,dx$

(ii) $\int_1^2 \frac{x}{(2+x)^3}\,dx$

(iii) $\int_0^3 (4x-1)\sqrt{2x^2-x+1}\,dx$

(iv) $\int_1^\infty \frac{x+1}{(x^2+2x)^5}\,dx$

(v) $\int_1^{\sqrt{2}} x^3(x^2-1)^6\,dx$

(vi) $\int_0^{\sqrt{3}} x^5 \sqrt{x^2+1}\,dx$

3 Use the substitution $u=1+\dfrac{1}{x}$ to evaluate $\displaystyle\int_1^2 \frac{\left(1+\dfrac{1}{x}\right)^3}{x^2}\,dx.$

4 The diagram shows the curve $y = x\sqrt{4+x}$.

(i) Find the co-ordinates of the point A.

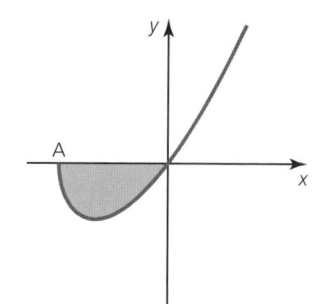

(ii) Find the area of the shaded region using the substitution $u = 4 + x$.

Substitution with exponentials and natural logarithms

EXERCISE 8.2

1 Find the integrals using the given substitutions.

(i) $\int 8xe^{x^2+3}dx$ $u = x^2 + 3$

(ii) $\int \dfrac{1-2x}{x-x^2}dx$ $u = x - x^2$

(iii) $\int \dfrac{e^x}{e^x - 1}dx$ $u = e^x - 1$

(iv) $\int \dfrac{e^{3x}}{(1+e^{3x})^3}dx$ $u = 1 + e^{3x}$

(v) $\int \dfrac{e^{\frac{1}{x}}}{x^2}dx$ $u = \dfrac{1}{x}$

(vi) $\int \dfrac{6x^2}{1-3x^3}dx$ $u = 1 - 3x^3$

(vii) $\int e^{1-\cos x} \sin x \, dx$ $u = 1 - \cos x$

2 Using the substitution $u = 2x + 1$, show that $\int_0^1 \dfrac{x}{2x+1} dx = \dfrac{1}{4}(2 - \ln 3)$.

OCR MEI Structured Mathematics C3 4753/1 May 2005 Q5

3 Find the exact value of $\int_0^{\ln 2} 32 e^{2x}(1 + e^{2x})^4 \, dx$ using the substitution $u = 1 + e^{2x}$.

4 Use the substitution $u = 2 + \ln t$ to find the exact value of $\int_1^e \dfrac{1}{t(2 + \ln t)^2} dt$.

5 The diagram shows part of the curve $y = \dfrac{e^{2x}}{1+e^{2x}}$.

The curve crosses the y axis at the point P.

(i) Find the co-ordinates of P.

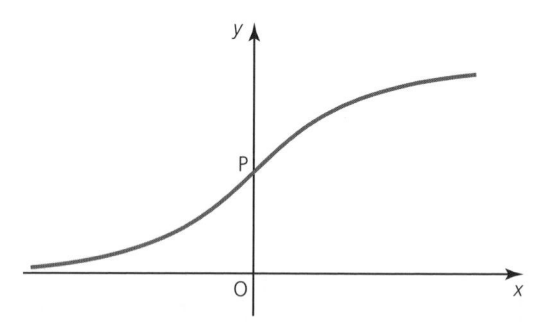

(ii) Find $\dfrac{dy}{dx}$, simplifying your answer.

Hence find the gradient of the curve at P.

(iii) Show that the area enclosed by the curve, the x axis, the y axis and the line $x = 1$ is $\dfrac{1}{2}\ln\left(\dfrac{1+e^2}{2}\right)$.

OCR MEI Mathematics C3 4753/1 June 2010 Q9

Integrals involving trigonometrical functions

EXERCISE 8.3

1 Find the following indefinite integrals using the given substitution.

(i) $\int x\sin(x^2)\,dx$ $\quad u=x^2$

(ii) $\int \cos 3x(1+\sin 3x)^3\,dx \quad u=1+\sin 3x$

(iii) $\int \sin x\, e^{\cos x}\,dx \quad u=\cos x$

(iv) $\int \cot x\,dx \quad u=\sin x$

(v) $\int \tan 4x\,dx \quad u=\cos 4x$

(vi) $\int \cos^5 x \sin x\,dx \quad u=\cos x$

(vii) $\int 2\sin^3 3x \cos 3x\,dx \quad u=\sin 3x$

(viii) $\int \dfrac{1}{\cos^2 2\theta}\,d\theta \quad u=2\theta$

2 Using the substitution $u = 1 - \cos x$, find $\int \sin x \cos x (1 - \cos x)^3 \, dx$.

3 Calculate $\int_0^1 \sqrt{4 - x^2} \, dx$ using the substitution $x = 2\sin\theta$.

4 Find the exact value of $\int_0^1 \dfrac{x^2}{\sqrt{1 - x^2}} \, dx$ using the substitution $x = \cos\theta$.

5 (i) Use the substitution $x = \sin^2\theta$ to show that $\int_0^1 \sqrt{\frac{1-x}{x}}\,dx = \int_0^{\frac{\pi}{2}} 2\cos^2\theta\,d\theta$.

(ii) Hence find the exact value of $\int_0^1 \sqrt{\frac{1-x}{x}}\,dx$.

6 Find $\int \frac{1}{x^2\sqrt{x^2-4}}\,dx$ using the substitution $x = 2\sec\theta$.

7 Use the substitution $x = \tan\theta$ to find the exact value of $\displaystyle\int_1^{\sqrt{3}} \frac{1-x^2}{1+x^2}\,dx$.

8 (i) Given that $x = \dfrac{1}{y}$, show that $\displaystyle\int \frac{1}{x\sqrt{x^2-1}}\,dx = -\int \frac{1}{\sqrt{1-y^2}}\,dy$.

(ii) Hence find $\displaystyle\int \frac{1}{x\sqrt{x^2-1}}\,dx$.

Partial fractions in integration

EXERCISE 8.4

1 Use partial fractions to find these integrals.

(i) $\displaystyle\int \frac{7x+8}{(x+2)(x-1)}\,dx$

(ii) $\displaystyle\int \frac{5x^2+12x+1}{(x+3)(x^2+1)}\,dx$

2 (i) Express $\dfrac{2x+1}{(x-3)^2}$ in the form $\dfrac{A}{x-3}+\dfrac{B}{(x-3)^2}$, where A and B are constants.

(ii) Hence find the exact value of $\displaystyle\int_4^{10} \frac{2x+1}{(x-3)^2}\,dx$, giving your answer in the form $a + b\ln c$, where a, b and c are integers.

3 (i) Express $\dfrac{2x^3 + 3x^2 + 9x + 12}{x^2 + 4}$ in the form $Ax + B + \dfrac{Cx + D}{x^2 + 4}$, where the values of the constants A, B, C and D are to be stated.

(ii) Use the result of part **(i)** to find the exact value of $\displaystyle\int_1^3 \dfrac{2x^3 + 3x^2 + 9x + 12}{x^2 + 4}\,dx$.

OCR Core Mathematics 4 4724/01 June 2007 Q7(ii), (iii)

4 Calculate the exact value of $\displaystyle\int_0^{\frac{\pi}{2}} \dfrac{\cos\theta}{\sin^2\theta - 5\sin\theta + 6}\,d\theta$ using the substitution $u = \sin\theta$.

5 (i) Show that the substitution $u = \sqrt{x}$ transforms $\int \dfrac{1}{x(1+\sqrt{x})}\,dx$ to $\int \dfrac{2}{u(1+u)}\,du$.

(ii) Hence find the exact value of $\displaystyle\int_1^9 \dfrac{1}{x(1+\sqrt{x})}\,dx$.

OCR Pure Mathematics 3 2633 January 2005 Q7

6 Let $I = \displaystyle\int_1^2 \dfrac{3}{x+\sqrt{2-x}}\,dx$.

(i) Using the substitution $u = \sqrt{2-x}$, show that $I = \displaystyle\int_0^1 \dfrac{6u}{(2-u)(1+u)}\,du$.

(ii) Hence show that $I = 2\ln 2$.

Integration by parts

EXERCISE 8.5

1 Use integration by parts to find the following integrals.

(i) $\int x \sin x \, dx$

(ii) $\int 4x e^x \, dx$

(iii) $\int x^2 \ln x \, dx$

2 Evaluate these definite integrals.

(i) $\int_1^2 (2x-1)e^{2x} \, dx$

(ii) $\int_0^{\frac{\pi}{6}} x \cos 2x \, dx$

(iii) $\int_1^3 2x \ln x \, dx$

3 The diagram shows part of the curve $y = x\cos 3x$.

The curve crosses the x axis at O, P and Q.

(i) Find the exact co-ordinates of P and Q.

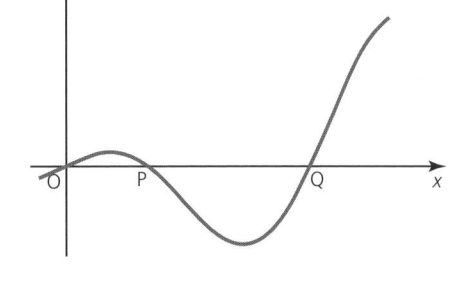

(ii) Find the exact gradient of the curve at the point P.

(iii) Find the area of the region enclosed by the curve and the x axis between O and P, giving your answer in exact form.

OCR MEI Mathematics C3 4753/1 January 2010 Q8

4 Find the exact value of $\int_1^e x^4 \ln x \, dx$.

5 (i) State the derivative of $e^{\cos x}$.

(ii) Hence use integration by parts to find the exact value of $\int_0^{\frac{1}{2}\pi} \cos x \sin x \, e^{\cos x} \, dx$.

6 Show that $\int_0^\pi (x^2 + 5x + 7)\sin x \, dx = \pi^2 + 5\pi + 10$.

7 (i) Differentiate $\dfrac{\ln x}{x^2}$, simplifying your answer.

(ii) Using integration by parts, show that $\displaystyle\int \dfrac{\ln x}{x^2}\,dx = -\dfrac{1}{x}(1+\ln x)+c$.

8 Let $I = \displaystyle\int e^x \sin x \, dx$.

Show, by using integration by parts, that $I = \dfrac{e^x(\sin x - \cos x)}{2}+c$.

Stretch and challenge

1 (i) Use the substitution $u = 1 + x$ to find the exact value of $\int_0^1 \dfrac{x^3}{1+x} \, dx$.

(ii) The diagram shows the curve $y = x^2 \ln(1+x)$.

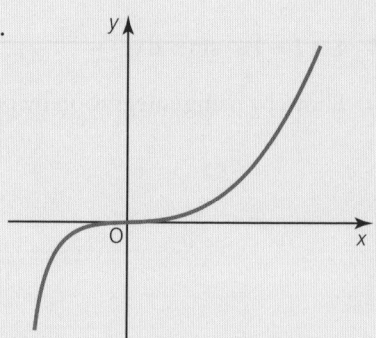

(a) Find $\dfrac{dy}{dx}$.

(b) Using integration by parts, and the result of part **(i)**, find the exact area enclosed by the curve $y = x^2 \ln(1+x)$, the x axis and the line $x = 1$.

OCR MEI Mathematics C3 4753/1 January 2011 Q8

2. It is given that, for non-negative integers n, $I_n = \int_0^{\frac{1}{2}\pi} x^n \cos x \, dx$.

   (i) Prove that, for $n \geq 2$, $I_n = \left(\frac{1}{2}\pi\right)^n - n(n-1)I_{n-2}$.

   (ii) Find I_4 in terms of π.

3. (i) Show that $\int_0^a f(x) \, dx = \int_0^a f(a-x) \, dx$.

(ii) Hence, or otherwise, calculate the value of the following integral, showing clearly the steps in your working:

$$\int_0^{\frac{\pi}{2}} \frac{\sin^n x}{\sin^n x + \cos^n x}\,dx, \text{ for any integer, } n.$$

NZQA Scholarship Calculus 2004 Q3(b)

4 (i) The function y is defined by $y = \dfrac{x^2}{1+x^2}$ $-1 \leqslant x \leqslant 1$.

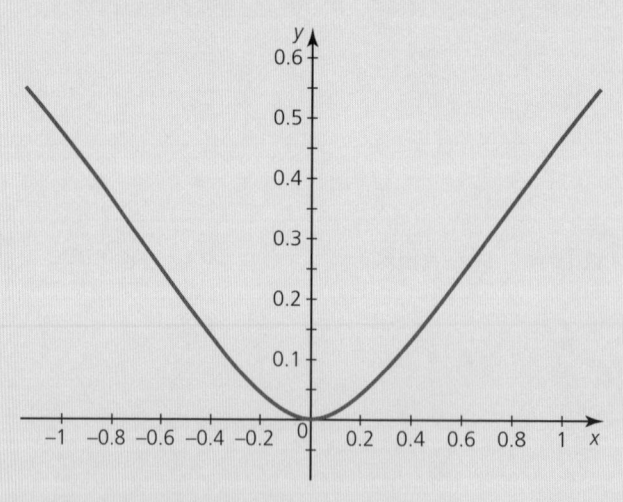

The gradient at the point $x = 1$ is $\frac{1}{2}$.

Hence show that there is a point with $\frac{1}{4} < x < \frac{1}{2}$, where the gradient is also $\frac{1}{2}$.

(ii) The shape of a wooden ornament is made by rotating the area between the graph of the function

$$y = \frac{(x-1)^2}{1+(x-1)^2} \quad 0 \leq x \leq 2$$

and the line $y = \frac{1}{2}$ through an angle 2π about the line $x = 1$.

Find the volume of this wooden ornament.

NZQA Scholarship Calculus 2004 Q3(a)

Exam focus

1 By first expressing $\dfrac{4x^2+4x-17}{2x^2+5x-3}$ in partial fractions, show that

$$\int_1^2 \dfrac{4x^2+4x-17}{2x^2+5x-3}\,dx = 2-\ln\dfrac{45}{4}.$$ [10]

2 (i) Show that $\dfrac{d}{dx}(\sec x) = \sec x \tan x$. [2]

(ii) Use your answer to part **(i)** to find $\int x^2 \sec x^3 \tan x^3 \,dx$, using the substitution $u = x^3$. [5]

(iii) Show that $\int_0^{\frac{\pi}{3}} (\sec x + \tan x)^2 \,dx = 2(\sqrt{3}+1) - \dfrac{\pi}{3}$. [5]

3 The diagram shows the curve $y = 4\sin^2 x \cos^3 x$.

(i) Find the x co-ordinate of the maximum point M. [5]

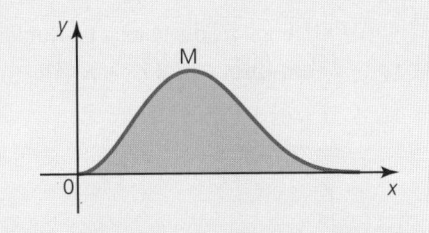

(ii) Using the substitution $u = \sin x$, find, by integration, the area of the shaded region bounded by the curve and the x axis. [5]

4 (i) Given that

$A(\sin\theta + \cos\theta) + B(\cos\theta - \sin\theta) \equiv 4\sin\theta,$

find the values of the constants A and B. [2]

(ii) Hence find the exact value of $\int_0^{\frac{1}{4}\pi} \frac{4\sin\theta}{\sin\theta+\cos\theta} d\theta$, giving your answer in the form $a\pi - \ln b$. [5]

5 Use the substitution $x = \sin\theta$ to find the exact value of $\int_0^{\frac{1}{2}} \frac{1}{(1-x^2)^{\frac{3}{2}}} dx$. [5]

6 The diagram shows the curve $y = x\sqrt{\ln x}$.

The shaded region between the curve, the x axis and the line $x = e$ is denoted by R.

(i) Find the equation of the tangent to the curve at the point where $x = 2$, giving your answer in the form $y = mx + c$. [4]

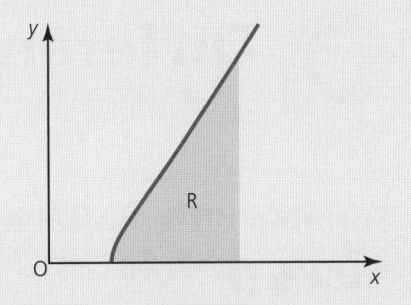

(ii) Find by integration the volume of the solid obtained when the region R is rotated completely about the x axis. Give your answer in terms of π and e. [7]

9 Differential equations

Forming differential equations

EXERCISE 9.1

1 Form a differential equation based on each situation described.

Do not solve the differential equations.

(i) The rate of change of the velocity of a car is constant.

(ii) The rate of growth of the number of bacteria is proportional to the number of bacteria.

(iii) The rate of change of the height of a tree is proportional to the cube root of the age of the tree in years.

(iv) The rate of change of the volume of water in a leaking tank is proportional to the square root of the volume of water left.

(v) The rate of change of the radius of a balloon as it is blown up is inversely proportional to the square root of the radius of the balloon.

(vi) The rate of change of the number of people hearing about a new product is proportional to the number of people who have not heard about the product.

(vii) The rate of change of the volume of water in a swimming pool due to evaporation is proportional to the surface area of the pool.

(viii) The area of a circular oil slick is increasing at a rate proportional to the square of the radius of the oil slick.

Solving differential equations

EXERCISE 9.2

1 Find the general solution to the following differential equations.

(i) $\dfrac{dy}{dx} = 1 - 2x$

(ii) $\dfrac{dy}{dx} = \dfrac{4x}{y}$

(iii) $\dfrac{dx}{dt} = xe^{2t}$

(iv) $\dfrac{dA}{dt} = 0.01A$

(v) $\dfrac{dy}{dx} = y - y\sin x$

(vi) $\dfrac{dy}{dt} = \dfrac{\cos^2 y}{e^t}$

2 Find the particular solution of these differential equations.

(i) $\dfrac{dy}{dx} = e^{x+y} \quad x = 0, \ y = 2$

(ii) $(1 + x^2)\dfrac{dy}{dx} = 2xy \quad x = 0, \ y = 2$

3 (i) Find the general solution of the differential equation $\dfrac{\sec^2 y}{\cos^2 2x}\dfrac{dy}{dx}=2$.

(ii) For the particular solution in which $y=\tfrac{1}{4}\pi$ when $x=0$, find the value of y when $x=\tfrac{1}{6}\pi$.

OCR Core Mathematics 4 4724/01 January 2007 Q9

4 The height, h metres, of a shrub t years after planting is given by the differential equation $\dfrac{dh}{dt}=\dfrac{6-h}{20}$.

A shrub is planted when its height is 1 m.

(i) Solve the differential equation to find an expression for t in terms of h.

(ii) How long after planting will the shrub reach a height of 2 m?

(iii) Find the height of the shrub 10 years after planting.

(iv) State the maximum possible height of the shrub.

OCR Core Mathematics 4 4724/01 June 2007 Q8

5 A liquid is being heated in an oven maintained at a constant temperature of 160 °C. It may be assumed that the rate of increase of the temperature of the liquid at any particular time, t minutes, is proportional to $160 - \theta$, where θ °C is the temperature of the liquid at that time.

When the liquid was placed in the oven, its temperature was 20 °C and 5 minutes later its temperature had risen to 65 °C.

Find the temperature of the liquid, correct to the nearest degree, after another 5 minutes.

6 (i) Find the general solution of the differential equation $\dfrac{dy}{dx} = \left(\dfrac{y}{x}\right)^2$, giving your answer in the form $y = f(x)$.

(ii) For the particular solution in which $y = 1$ when $x = 2$, find the value of y when $x = 8$.

OCR Pure Mathematics 3 2633 January 2005 Q6

7 (i) Find the quotient and the remainder when $x^2 - 5x + 6$ is divided by $x - 1$.

(ii) (a) Find the general solution of the differential equation

$$\left(\frac{x-1}{x^2 - 5x + 6}\right)\frac{dy}{dx} = y - 5.$$

(b) Given that $y = 7$ when $x = 8$, find y when $x = 6$.

8 Paraffin is stored in a tank with a horizontal base. At time t minutes the depth of paraffin in the tank is x cm. When $t = 0$, $x = 72$.

There is a tap in the side of the tank through which the paraffin can flow. When the tap is opened the flow of the paraffin is modelled by the differential equation
$$\frac{dx}{dt} = -4(x-8)^{\frac{1}{3}}.$$

(i) How long does it take for the level of paraffin to fall from a depth of 72 cm to a depth of 35 cm?

(ii) The tank is filled again to its original depth of 72 cm of paraffin and the tap is then opened. The paraffin flows out until it stops. How long does this take?

9 (i) The number of bacteria in a colony is increasing at a rate that is proportional to the square root of the number of bacteria present.

Form a differential equation relating the number of bacteria, x, to the time, t.

(ii) In another colony the number of bacteria, y, after time t minutes is modelled by the differential equation $\dfrac{dy}{dt} = \dfrac{10000}{\sqrt{y}}$.

Find y in terms of t, given that $y = 900$ when $t = 0$. Hence find the number of bacteria after 10 minutes.

OCR MEI Structured Mathematics C4 4754(A) June 2006 Q4

10 (i) Express $\dfrac{1}{(2x+1)(x+1)}$ in partial fractions.

(ii) A curve passes through the point $(0, 2)$ and satisfies the differential equation
$$\dfrac{dy}{dx} = \dfrac{y}{(2x+1)(x+1)}.$$
Show by integration that $y = \dfrac{4x+2}{x+1}$.

OCR MEI Applications of Advanced Mathematics C4 4754(A)/01 January 2007 Q6

11 A skydiver drops from a helicopter. Before she opens her parachute, her speed v ms^{-1} after time t seconds is modelled by the differential equation $\dfrac{dv}{dt} = 10e^{-\frac{1}{2}t}$.

When $t = 0$, $v = 0$.

(i) Find v in terms of t.

(ii) According to this model, what is the speed of the skydiver in the long term?

She opens her parachute when her speed is 10 ms^{-1}. Her speed t seconds after this is w ms^{-1}, and is modelled by the differential equation $\dfrac{dw}{dt} = -\dfrac{1}{2}(w-4)(w+5)$.

(iii) Express $\dfrac{1}{(w-4)(w+5)}$ in partial fractions.

(iv) Using this result, show that $\dfrac{w-4}{w+5} = 0.4e^{-4.5t}$.

(v) According to this model, what is the speed of the skydiver in the long term?

OCR MEI Applications of Advanced Mathematics C4 4754/01A May 2008 Q9

12 The forensic team at C.S.I. discover a dead body at 4pm in a shallow stream of water. At that time the man's temperature is 16 °C and the temperature of the water is 6 °C. One hour later, the man's temperature has dropped to 8 °C.

Can you help the team determine the time the man died?

(The team understands Newton's law of cooling which states that the rate of change of temperature of an object is proportional to the difference between the temperature of the object and the surrounding temperature. The normal human body temperature is 37 °C.)

13 A cylindrical water tank has developed a leak at the bottom of the tank. It is known that the rate that water leaks from the tank is proportional to the square root of the volume of water remaining.

If the tank had 900 litres in it initially and after one day 59 litres were lost due to the leak, how long will it take for the tank to empty?

14 During a chemical reaction, substance A is converted into substance B at a rate that is proportional to the square of the amount of A.

When $t = 0$, 60 grams of A are present, and after 1 hour ($t = 1$), only 10 grams of A remain unconverted.

How much of A is present after 2 hours?

Stretch and challenge

1 Some years ago an island was populated by red squirrels and there were no grey squirrels. Then grey squirrels were introduced.

The population x, in thousands, of red squirrels is modelled by the equation $x = \dfrac{a}{1+kt}$, where t is the time in years, and a and k are constants.

When $t = 0$, $x = 2.5$.

(i) Show that $\dfrac{dx}{dt} = -\dfrac{kx^2}{a}$.

(ii) Given that the initial population of 2.5 thousand red squirrels reduces to 1.6 thousand after one year, calculate a and k.

(iii) What is the long-term population of red squirrels predicted by this model?

The population y, in thousands, of grey squirrels is modelled by the differential equation $\dfrac{dy}{dt} = 2y - y^2$.

When $t = 0$, $y = 1$.

(iv) Express $\dfrac{1}{2y - y^2}$ in partial fractions.

(v) Hence show by integration that $\ln\left(\dfrac{y}{2-y}\right) = 2t$.

(vi) What is the long-term population of grey squirrels predicted by this model?

OCR MEI Structured Mathematics C4 4754(A) January 2006 Q8

2 In a bush reserve the number of possums, P, is given by the formula $P = \dfrac{700}{5 + 2e^{-\frac{t}{2}}}$, where t is time in years from today.

(i) How many possums are there in the reserve today?

(ii) Find the long-term number of possums that this model predicts.

(iii) By expressing $e^{-\frac{t}{2}}$ in terms of P, show that $\dfrac{dP}{dt} = \dfrac{P}{2}\left(1 - \dfrac{P}{140}\right)$.

(iv) Hence find the instantaneous yearly rate of change of the number of possums in the reserve today.

3 Find **all** functions $y = f(x)$ which satisfy $\dfrac{dy}{dx} = y^{m+1}$, where m is a non-zero constant. Show that these functions also satisfy $\dfrac{d}{dx}(y^n) = ny^{n+m}$.

NZQA Scholarship Calculus 2011 Q1(c)

4 (i) During a penalty conversion attempt a rugby ball may be considered to be a particle that moves as a projectile. Hence you may ignore the effects of air resistance, etc. Initially, before the ball is kicked, the particle is considered to be at the origin, O, relative to a horizontal x axis and a perpendicular y axis. Assume that the particle moves only in the plane of the x–y axes.

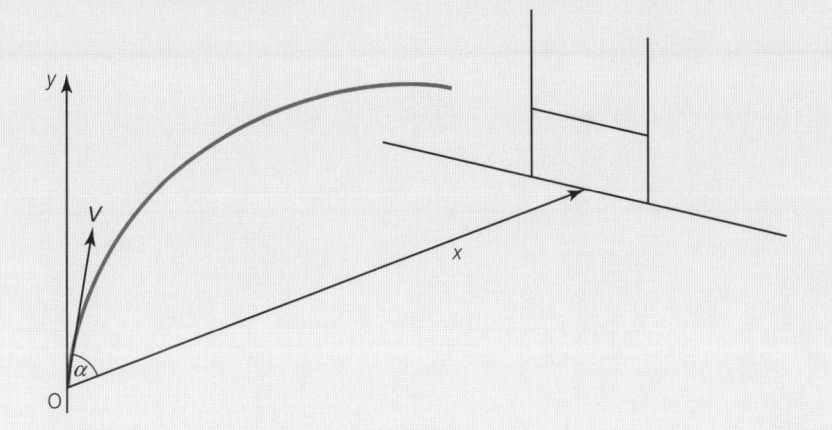

The acceleration of the particle after the ball has been kicked, measured relative to these two axes, is given by:

$$\frac{d^2x}{dt^2} = 0 \quad \text{and} \quad \frac{d^2y}{dt^2} = -g$$

where g is the acceleration due to gravity.

The particle leaves the ground with a speed of V metres per second at an angle of α to the x axis. By using integration, or otherwise, find the equation of the path of the particle.

(ii) On a particular conversion attempt the co-ordinates of the centre of the goalpost crossbar are (kh, h).

(a) Show that there are two possible paths by which the particle may hit the centre of the crossbar if $V^2 > gh\left(1 + \sqrt{1+k^2}\right)$.

(b) In this case, show that for these two possible angles α_1 and α_2,
$\alpha_1 + \alpha_2 = \tan^{-1}(-k)$.

NZQA Scholarship Calculus 2004 Q6

5. The typical population growth model assumes the rate of growth of the population is proportional to the size of the population. Other models take into account factors that limit the growth. This question describes a type of growth model called a **Gompertz growth model**. This model assumes that the rate of change of y is proportional to y and the natural log of $\frac{L}{y}$, where L is the population limit.

A population of 10 tigers has been introduced into a national park. The forest service estimates that the maximum population the park can sustain is 180 tigers. After 2 years, the population is estimated to be 30 tigers. If the population follows a Gompertz growth model, how many tigers will there be 12 years after their introduction?

6 A tank contains 40 litres of a solution composed of 80% water and 20% alcohol. A second solution containing half water and half alcohol is added to the tank at the rate of 4 litres per minute. At the same time, the tank is being drained at the rate of 4 litres per minute. Assuming that the solution is stirred constantly, how much alcohol will be in the tank after 15 minutes?

Exam focus

1 (i) Show that, if $y = \operatorname{cosec} x$, then $\dfrac{dy}{dx}$ can be expressed as $-\operatorname{cosec} x \cot x$. [3]

(ii) Solve the differential equation $\dfrac{dx}{dt} = -\sin x \tan x \cot t$, given that $x = \tfrac{1}{6}$ when $t = \tfrac{1}{2}$. [4]

2 (i) Express $\dfrac{1}{(3-x)(6-x)}$ in partial fractions. [2]

(ii) In a chemical reaction the amount, x grams, of a substance present at time t seconds is related to the rate at which x is changing by the equation

$\dfrac{dx}{dt} = k(3-x)(6-x)$, where k is a constant.

When $t = 0$, $x = 0$ and when $t = 1$, $x = 1$.

(a) Show that $k = \tfrac{1}{3}\ln\tfrac{5}{4}$. [6]

(b) Find the value of x when $t = 2$. [3]

3 Solve the differential equation $e^{2y}\dfrac{dy}{dx} + \tan x = 0$ given that when $x = 0$, $y = 0$.

Give your answer in the form $y = f(x)$. [6]

4 A biologist is investigating the spread of algae in a lake. At time t weeks after the start of the investigation, the area covered by the weed is $A\,\text{m}^2$. The biologist claims that the rate of increase of A is proportional to $\sqrt{3A-2}$.

    (i) Write down a differential equation representing the biologist's claim. [1]

    (ii) At the start of the investigation, the area covered by the algae was $6\,\text{m}^2$ and 4 weeks later the area covered was $17\,\text{m}^2$.

    Assuming that the biologist's claim is correct, find the area covered 8 weeks after the start of the investigation. [9]

5 Given that $y = 0$ when $x = 2$, solve the differential equation $xy^2 \dfrac{dy}{dx} = y^3 - 1$ obtaining an expression for y^3 in terms of x. [6]

10 Vectors

The vector equation of a line

EXERCISE 10.1

1 Find an equation of the following lines in vector form.

  (i) The line going through the point (3, −1) parallel to the vector −2**i** + 5**j**.

  (ii) The line passing through the points A(−2, 1) and B(0, 8).

  (iii) The line passing through the point with position vector 4**i** − 3**j** + **k** in the same direction as the vector **i** − 4**k**.

(iv) The line passing through the points A(1, 9, −5) and B(2, 0, −1).

(v) The line passing through the points C(2, 3, 0) and D(−3, −4, 5).

(vi) The line going through the point with position vector 2**i** + 3**j** − 2**k** parallel to $\begin{pmatrix} 8 \\ -2 \\ 1 \end{pmatrix}$.

(vii) The line going through A, parallel to DC (where points A, C and D are those given in parts **(iv)** and **(v)**).

2 Determine if the following points lie on the line given.

(i) (–7, –7, 4) $\quad \mathbf{r} = \begin{pmatrix} 2 \\ -1 \\ 1 \end{pmatrix} + \lambda \begin{pmatrix} -3 \\ -2 \\ 1 \end{pmatrix}$

(ii) (0, 2, –8) $\quad \mathbf{r} = \begin{pmatrix} 1 \\ 0 \\ -6 \end{pmatrix} + \mu \begin{pmatrix} 1 \\ -2 \\ 1 \end{pmatrix}$

3 The straight line l has the vector equation $\mathbf{r} = 3\mathbf{i} - 4\mathbf{j} + 2\mathbf{k} + t(\mathbf{i} + 3\mathbf{j} - \mathbf{k})$.

Given that the point $(a, b, 0)$ lies on the line, find the values of a and b.

The intersection of two lines

EXERCISE 10.2

1 Find the point of intersection of the two lines given.

If the lines do not intersect, state whether they are parallel or skew.

(i) $\mathbf{r} = \begin{pmatrix} 4 \\ -1 \\ 1 \end{pmatrix} + \lambda \begin{pmatrix} 2 \\ -2 \\ 3 \end{pmatrix}$ and $\mathbf{r} = \begin{pmatrix} 3 \\ -3 \\ 7 \end{pmatrix} + \mu \begin{pmatrix} -5 \\ 2 \\ 0 \end{pmatrix}$.

(ii) $\mathbf{r} = \mathbf{i} - \mathbf{j} + 3\mathbf{k} + s(2\mathbf{i} + 4\mathbf{j} + \mathbf{k})$ and $\mathbf{r} = 5\mathbf{i} + 2\mathbf{j} + \mathbf{k} + t(-\mathbf{i} + 3\mathbf{j} + 2\mathbf{k})$.

(iii) $\mathbf{r} = \begin{pmatrix} 1-2\lambda \\ 3\lambda \\ 2+\lambda \end{pmatrix}$ and $\mathbf{r} = \begin{pmatrix} -1+4\mu \\ 1-5\mu \\ -3+\mu \end{pmatrix}$.

(iv) $\mathbf{r} = \begin{pmatrix} 1 \\ 8 \\ 0 \end{pmatrix} + \lambda \begin{pmatrix} 4 \\ -1 \\ 2 \end{pmatrix}$ and $\mathbf{r} = \begin{pmatrix} 1 \\ 0 \\ 12 \end{pmatrix} + \mu \begin{pmatrix} -8 \\ 2 \\ 4 \end{pmatrix}$.

(v) $\mathbf{r} = (-4 - s)\mathbf{i} + 4\mathbf{j} + (-5 + 3s)\mathbf{k}$ and $\mathbf{r} = (3 + 11t)\mathbf{i} + (1 - 3t)\mathbf{j} + (8 + t)\mathbf{k}$.

2 The vector equations of two lines are

$\mathbf{r} = (5\mathbf{i} - 2\mathbf{j} - 2\mathbf{k}) + s(3\mathbf{i} - 4\mathbf{j} + 2\mathbf{k})$ and $\mathbf{r} = (2\mathbf{i} - 2\mathbf{j} + 7\mathbf{k}) + t(2\mathbf{i} - \mathbf{j} - 5\mathbf{k})$.

Prove that the two lines are:

(i) perpendicular

(ii) skew.

3 The equations of two lines are given by

$\mathbf{r} = \begin{pmatrix} 1 \\ 2 \\ -1 \end{pmatrix} + \lambda \begin{pmatrix} -1 \\ 2 \\ 3 \end{pmatrix}$ and $\mathbf{r} = \begin{pmatrix} 0 \\ 6 \\ 3 \end{pmatrix} + \mu \begin{pmatrix} 1 \\ 0 \\ -2 \end{pmatrix}$.

Find the point of intersection of the lines.

4 Show that the straight line with equation $\mathbf{r} = \begin{pmatrix} 2 \\ -3 \\ 5 \end{pmatrix} + t \begin{pmatrix} 1 \\ 4 \\ -2 \end{pmatrix}$ meets the line passing through (9, 7, 5) and (7, 8, 2), and find the point of intersection of these lines.

5 Relative to an origin O, the points A and B have position vectors $5\mathbf{i} + \mathbf{j} + 3\mathbf{k}$ and $7\mathbf{i} + 4\mathbf{k}$ respectively.

(i) Find a vector equation of the line passing through A and B.

(ii) Find the position vector of the point P on AB such that OP is perpendicular to AB.

6 The points P, Q, R, and S have position vectors, relative to the origin O, given by the following.

$$\vec{OP} = \begin{pmatrix} 3 \\ 4 \\ 7 \end{pmatrix} \quad \vec{OQ} = \begin{pmatrix} 13 \\ 9 \\ 2 \end{pmatrix} \quad \vec{OR} = \begin{pmatrix} 1 \\ 2 \\ 3 \end{pmatrix} \quad \vec{OS} = \begin{pmatrix} 10 \\ k \\ 6 \end{pmatrix}$$

The lines PQ and RS intersect at the point A.

(i) Find the value of k.

(ii) Find the co-ordinates of A.

7 Lines l_1 and l_2 have vector equations $\mathbf{r} = \mathbf{j} + \mathbf{k} + t(2\mathbf{i} + a\mathbf{j} + \mathbf{k})$ and $\mathbf{r} = 3\mathbf{i} - \mathbf{k} + s(2\mathbf{i} + 2\mathbf{j} - 6\mathbf{k})$ respectively, where t and s are parameters and a is a constant.

(i) Given that l_1 and l_2 are perpendicular, find the value of a.

(ii) Given instead that l_1 and l_2 intersect, find the value of a.

8 The lines l and m have the following equations.

$\mathbf{r} = -3\mathbf{i} + 2\mathbf{j} + 3\mathbf{k} + \lambda(\mathbf{i} + 2\mathbf{j} + \mathbf{k})$ and $\mathbf{r} = 7\mathbf{i} + 3\mathbf{j} + 3\mathbf{k} + \mu(a\mathbf{i} + b\mathbf{j} - 2\mathbf{k})$

(i) Given that l and m intersect, show that $a - 10b = 38$.

(ii) Given also that l and m are perpendicular, find the values of a and b.

(iii) When a and b have these values, find the position vector of the point of intersection of l and m.

The angle between two lines

EXERCISE 10.3

1 The equations of two lines are given by

$$\mathbf{r} = \begin{pmatrix} 1 \\ 2 \\ -1 \end{pmatrix} + \lambda \begin{pmatrix} -1 \\ 2 \\ 3 \end{pmatrix} \quad \text{and} \quad \mathbf{r} = \begin{pmatrix} 0 \\ 6 \\ 3 \end{pmatrix} + \mu \begin{pmatrix} 1 \\ 0 \\ -2 \end{pmatrix}.$$

Find the acute angle between the lines.

2 (i) The vector $\mathbf{u} = 3\mathbf{i} + b\mathbf{j} + c\mathbf{k}$ is perpendicular to the vector $4\mathbf{i} + \mathbf{k}$ and to the vector $4\mathbf{i} + 3\mathbf{j} + 2\mathbf{k}$.

Find the values of b and c.

(ii) Calculate, to the nearest degree, the angle between the vectors $4\mathbf{i} + \mathbf{k}$ and $4\mathbf{i} + 3\mathbf{j} + 2\mathbf{k}$.

3 Find the acute angle between the line with equation

$\mathbf{r} = (2\mathbf{i} - \mathbf{j} - 3\mathbf{k}) + s(3\mathbf{i} - 5\mathbf{j} + 2\mathbf{k})$

and the line passing through the points (9, 7, 5) and (7, 8, 2).

4 The co-ordinates of the vertices of a triangle are given by

A(1, 3, –2), B(8, –2, 4) and C(5, –1, a)

where a is a constant.

Find the value(s) of a such that angle ACB is 90°.

5 The diagram shows a square-based pyramid OABCD with base lengths of 1 unit and a height of 1 unit. The point D lies directly above the midpoint of the diagonal OB.

Find the angle between the vectors $\overrightarrow{OD}$ and $\overrightarrow{OB}$.

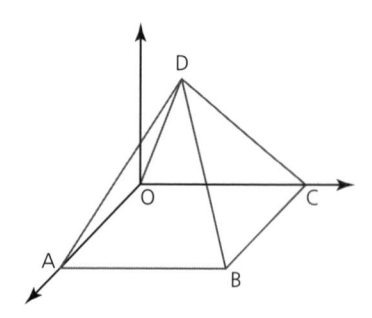

The perpendicular distance from a point to a line

EXERCISE 10.4

1 Find the shortest distance from the given point to the given straight line.

In each case find the co-ordinates of the point on the line that gives this minimum distance.

(i) A(0, −7, −3) $\mathbf{r} = \begin{pmatrix} 1 \\ -1 \\ 1 \end{pmatrix} + t \begin{pmatrix} -3 \\ 4 \\ 2 \end{pmatrix}$

(ii) B(−1, −6, 7) $\mathbf{r} = (4\mathbf{i} - 2\mathbf{j} + \mathbf{k}) + \lambda(2\mathbf{i} - \mathbf{j} - 2\mathbf{k})$

(iii) C(1, 1, −2) $\mathbf{r} = \begin{pmatrix} 0 \\ 1 \\ 3 \end{pmatrix} + s \begin{pmatrix} 5 \\ -1 \\ -2 \end{pmatrix}$

2 The equation of a straight line l is $\mathbf{r} = \begin{pmatrix} 3 \\ 1 \\ 1 \end{pmatrix} + t \begin{pmatrix} 1 \\ -1 \\ 2 \end{pmatrix}$. O is the origin.

   (i) The point P on l is given by $t = 1$.

   Calculate the acute angle between OP and l.

   (ii) Find the position vector of the point Q on l such that OQ is perpendicular to l.

   (iii) Find the length of OQ.

3 The position vectors of the points P and Q with respect to an origin O are $5\mathbf{i} + 2\mathbf{j} - 9\mathbf{k}$ and $4\mathbf{i} + 4\mathbf{j} - 6\mathbf{k}$ respectively.

(i) Find a vector equation for the line PQ.

The position vector of the point T is $\mathbf{i} + 2\mathbf{j} - \mathbf{k}$.

(ii) Write down a vector equation for the line OT and show that OT is perpendicular to PQ.

It is given that OT intersects PQ.

(iii) Find the position vector of the point of intersection of OT and PQ.

(iv) Hence find the perpendicular distance from O to PQ, giving your answer in an exact form.

4. A plane takes off at 10am from the point (1.2, 0.8, 0) and heads in the direction $-2\mathbf{i} - \mathbf{j} + \mathbf{k}$. All units are in kilometres. The control tower is located at the origin.

   (i) Find the closest distance the plane gets to the control tower.

   (ii) Find the height of the plane above the ground at this time.

5 An eagle, initially located at the point with position vector $-10\mathbf{i} + 20\mathbf{j} + 50\mathbf{k}$, spots a rabbit on the ground at the point with position vector $5\mathbf{i} + 20\mathbf{j}$.

Immediately, the eagle swoops down with velocity $5\mathbf{i} + \mathbf{j} - 10\mathbf{k}$ and the rabbit takes off with velocity $2\mathbf{i} + \mathbf{j}$.

The rabbit's burrow is located at $25\mathbf{i} + 30\mathbf{j}$. All the units are metres.

(i) How long would the rabbit take to reach its burrow travelling at $2\mathbf{i} + \mathbf{j}$ metres per second?

(ii) Determine if the eagle catches the rabbit.

The vector equation of a plane

EXERCISE 10.5

1. Write the following planes in cartesian form $ax + by + cx = d$.

   (i) $\mathbf{r} \cdot (2\mathbf{i} + 4\mathbf{j} - \mathbf{k}) = 4$

   (ii) $(\mathbf{r} - 4\mathbf{j}) \cdot (3\mathbf{i} - 2\mathbf{j} - \mathbf{k}) = 9$

   (iii) $\mathbf{r} = 3\mathbf{i} + \mathbf{j} - \mathbf{k} + \lambda(2\mathbf{i} + \mathbf{k}) + \mu(5\mathbf{i} + 2\mathbf{j} + 3\mathbf{k})$

2. Write the equation of the following planes in scalar product form $\mathbf{r} \cdot \mathbf{n} = k$.

   (i) $5x - 2y + z = 10$

   (ii) $\mathbf{r} = \mathbf{i} + 2\mathbf{j} - 2\mathbf{k} + \lambda(\mathbf{i} + 3\mathbf{j}) + \mu(2\mathbf{i} - \mathbf{j} + \mathbf{k})$

3 Find the distance of the following planes from the origin.

(i) $x - 5y - 3z = 70$

(ii) $\mathbf{r}.(5\mathbf{i} - 2\mathbf{j} - 3\mathbf{k}) = 19$

4 Find the equation of the following planes in cartesian form $ax + by + cz = d$ and scalar product form $\mathbf{n}.\mathbf{r} = k$.

(i) The plane which is perpendicular to the vector $2\mathbf{i} + \mathbf{j} - 3\mathbf{k}$ and contains the point $(3, -4, -1)$.

(ii) The plane which goes through the origin and is perpendicular to the vector $\mathbf{i} + 4\mathbf{j} + 5\mathbf{k}$.

(iii) The plane which is parallel to the plane $3x - y - z = 12$ and contains the point $(-1, 4, 1)$.

(iv) The plane which contains the lines $\mathbf{r} = (5\mathbf{i} - 2\mathbf{j} + 2\mathbf{k}) + s(3\mathbf{i} - 4\mathbf{j} + 2\mathbf{k})$ and $\mathbf{r} = (5\mathbf{i} + 2\mathbf{j} + 3\mathbf{k}) + t(-2\mathbf{i} + 4\mathbf{j} - \mathbf{k})$.

(v) The plane which contains the point (3, −1, 0) and the line
$\mathbf{r} = (4\mathbf{i} + \mathbf{j} + 2\mathbf{k}) + t(\mathbf{i} - \mathbf{j} - 3\mathbf{k})$.

(vi) The plane which contains the points A(2, 3, −1), B(−8, −2, 1) and C(1, −1, 6).

(vii) The plane which is perpendicular to the plane $2x - y + 3z = 5$ and contains the points (1, −1, 2) and (4, 0, 1).

5 Write down a vector equation of the line through (2, 0, 2) perpendicular to the plane $2x - y + z = 0$.

Find the point of intersection of this line with the plane.

6 In each case, determine whether the given line is parallel to the plane, lies entirely in the plane or intersects the plane at one point. If it does intersect at one point, find the co-ordinates of the point of intersection.

(i) $\mathbf{r} = \begin{pmatrix} 0 \\ -2 \\ 3 \end{pmatrix} + t \begin{pmatrix} 2 \\ -2 \\ -1 \end{pmatrix}$ and $2x + y + 2z = 8$

(ii) $\mathbf{r} = \begin{pmatrix} 3 \\ 2 \\ -1 \end{pmatrix} + s \begin{pmatrix} 4 \\ 2 \\ -3 \end{pmatrix}$ and $\mathbf{r} \cdot \begin{pmatrix} 1 \\ 2 \\ -3 \end{pmatrix} = -7$

(iii) $\mathbf{r} = (\mathbf{i} - \mathbf{j} + 2\mathbf{k}) + t(3\mathbf{i} - \mathbf{j} + \mathbf{k})$ and $x + y - 2z = -4$

(iv) $\mathbf{r} = (4 - s)\mathbf{i} + 3\mathbf{j} + (-2 + s)\mathbf{k}$ and $\mathbf{r} \cdot \begin{pmatrix} -1 \\ 4 \\ 3 \end{pmatrix} = 10$

7 Find the angle between the given line and plane.

(i) $\mathbf{r} = \begin{pmatrix} 0 \\ -2 \\ 3 \end{pmatrix} + t \begin{pmatrix} 2 \\ -2 \\ -1 \end{pmatrix}$ and $2x + y + z = 8$

(ii) $\mathbf{r} = (2+s)\mathbf{i} + (3-4s)\mathbf{j} + (1+s)\mathbf{k}$ and $\mathbf{r} \cdot \begin{pmatrix} 3 \\ 1 \\ -2 \end{pmatrix} = 12$

(iii) $\mathbf{r} = (-2\mathbf{i} - 3\mathbf{j} + \mathbf{k}) + t(2\mathbf{i} + \mathbf{j} + 3\mathbf{k})$ and $5x + y + 3z = 8$

8 In each of the following find the co-ordinates of the foot of the perpendicular from the point to the given plane and the minimum distance from the point to the plane.

(i) $(2, -3, 4)$ and $x + 2y + 2z = 13$

(ii) (2, 1, 1) and $\mathbf{r} \cdot \begin{pmatrix} 1 \\ 2 \\ -2 \end{pmatrix} = 20$

(iii) (−2, 1, −1) and $x - 2y + 3z = 0$

9 The line l has equation $\mathbf{r} = \begin{pmatrix} 1 \\ 3 \\ -2 \end{pmatrix} + \lambda \begin{pmatrix} 2 \\ 3 \\ 5 \end{pmatrix}$ and the plane p has equation $\mathbf{r} \cdot \begin{pmatrix} 3 \\ -7 \\ 3 \end{pmatrix} = k$.

(i) Given that l lies entirely in p, find the value of k.

(ii) Find the co-ordinates of the point on p which is closest to the point A(4, 10, −11). Find the shortest distance from p to A.

(iii) Find, in the form $ax + by + cz = d$, where a, b, c and d are integers, the equation of the plane which contains both l and A.

The intersection of two planes

EXERCISE 10.6

1 For each of the following find the equation of the line of intersection of the given planes and the acute angle between the given planes.

(i) $2x + z = 4$ and $2y - z = 6$

(ii) $\mathbf{r} \cdot \begin{pmatrix} 2 \\ -3 \\ 1 \end{pmatrix} = 6$ and $\mathbf{r} \cdot \begin{pmatrix} 1 \\ 2 \\ -5 \end{pmatrix} = -4$

(iii) $3x - y + 2z = 6$ and $x + 2y - 4z = 9$

2 The equations of two planes are given by

$p: x + 3y - 2z = 5$ and $\pi: -x - 3y + 2z = 1$.

(i) Show that the planes are parallel.

(ii) Show that the point A(5, 2, 3) lies on p and the point B(4, −3, −2) lies on π.

(iii) Find the vector $\overrightarrow{AB}$ and the length from A to B.

(iv) Find the angle between the normal to the planes and $\overrightarrow{AB}$.

(v) Hence find the distance between the planes.

(vi) The normal to the planes through A intersects the plane π at the point C. Find the co-ordinates of the point C.

(vii) Find the distance between A and C. (This should be the same as your answer to part **(v)**.)

3 The planes π_1 and π_2 have equations $3x - 2y + z = k$ and $\mathbf{r} \cdot \begin{pmatrix} 6 \\ 2 \\ -1 \end{pmatrix} = 18$ respectively.

The line l is defined by $\mathbf{r} = \begin{pmatrix} 1 \\ 2 \\ -3 \end{pmatrix} + \lambda \begin{pmatrix} -1 \\ 2 \\ 3 \end{pmatrix}$ and the point A is (4, 2, 0).

(i) The distance of the plane π_1 from the origin is $\dfrac{\sqrt{14}}{2}$. Find the value of k.

(ii) Find the acute angle between π_1 and π_2.

(iii) Calculate the shortest distance between A and the plane π_2.

(iv) Find the position vector of the point of intersection of l and π_2.

(v) Find the equation of the plane containing the line l and the point A.

4 Three points have co-ordinates A(1, −8, −2), B(9, 2, 4) and C(−3, 2, 10).

The plane p contains the point B and is perpendicular to the line AB.

The plane q contains the point C and is perpendicular to AC.

Find the equation of the line of intersection of p and q.

5 Find the distance between the two planes $2x - y + 2z + 4 = 0$ and $2x - y + 2z + 16 = 0$.

Find also the equation of the plane that is equidistant from both planes.

6 The planes $ax + 3y + z = k$ and $x + by + 2z = d$ intersect along the line with equation

$$\mathbf{r} = \begin{pmatrix} 2 \\ -1 \\ 0 \end{pmatrix} + \lambda \begin{pmatrix} 9 \\ -7 \\ c \end{pmatrix}.$$

Find the values of the constants a, b, c, d and k.

7 The diagram represents a house. All units are in metres. The co-ordinates of A, B, C and E are as shown. BD is horizontal and parallel to AE.

(i) Find the length AE.

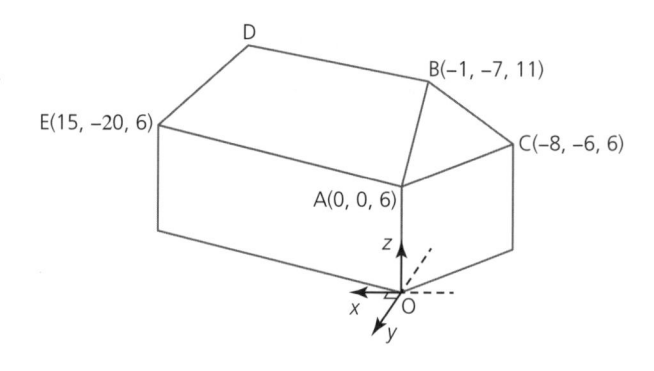

(ii) Find a vector equation of the line BD. Given that the length of BD is 15 metres, find the co-ordinates of D.

(iii) Verify that the equation of the plane ABC is $-3x + 4y + 5z = 30$.

Write down a vector normal to this plane.

(iv) Show that the vector $\begin{pmatrix} 4 \\ 3 \\ 5 \end{pmatrix}$ is normal to the plane ABDE.

Hence find the equation of the plane ABDE.

(v) Find the angle between the planes ABC and ABDE.

OCR MEI Structured Mathematics C4 4754(A) June 2006 Q7

8 The point A(–1, 12, 5) lies on the plane P with equation $8x - 3y + 10z = 6$.

The point B(6, –2, 9) lies on the plane Q with equation $3x - 4y - 2z = 8$.

The planes P and Q intersect in the line L.

(i) Find an equation for the line L.

The lines *M* and *N* are both parallel to *L*, with *M* passing through A and *N* passing through B.

(ii) Find the distance between the parallel lines *M* and *N*.

The point C has co-ordinates $(k, 0, 2)$, and the line AC intersects the line *N* at the point D.

(iii) Find the value of *k*, and the co-ordinates of D.

OCR MEI Further Applications of Advanced Mathematics FP3 4757 June 2009 Q1(i), (iii), (iv)

Stretch and challenge

1 Two mineshafts follow straight-line paths given by the equations

$$\mathbf{r} = \begin{pmatrix} -1 \\ -2 \\ -1 \end{pmatrix} + t \begin{pmatrix} 0 \\ -1 \\ -1 \end{pmatrix} \quad \text{and} \quad \mathbf{r} = \begin{pmatrix} -2 \\ -4 \\ -1 \end{pmatrix} + t \begin{pmatrix} -1 \\ -3 \\ -3 \end{pmatrix}.$$

The units are in kilometres.

A vertical ventilation shaft needs to be constructed at the point where the distance between the mineshafts is as small as possible.

(i) Find the co-ordinates of the points in both mineshafts where the shaft will be constructed.

(ii) Find the length of the ventilation shaft.

2 A computer-controlled machine can be programmed to make cuts by entering the equation of the plane of the cut and to drill holes by entering the equation of the line of the hole.

A 20 cm × 30 cm × 30 cm cuboid is to be cut and drilled. The cuboid is positioned relative to the x, y and z axes as shown.

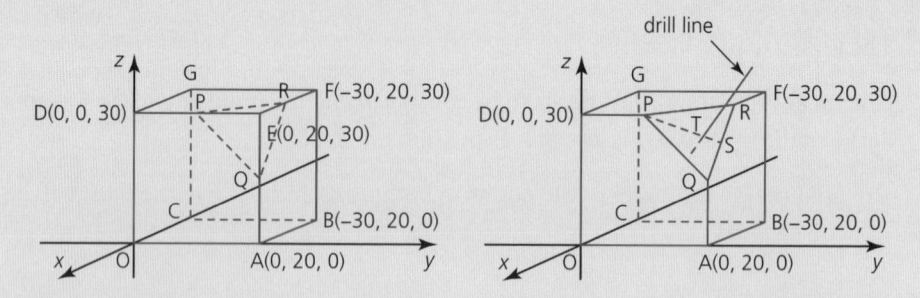

First, a plane cut is made to remove the corner at E. The cut goes through the points P, Q and R, which are the midpoints of the sides ED, EA and EF respectively.

(i) Write down the co-ordinates of P, Q and R.

Hence show that $\overrightarrow{PQ} = \begin{pmatrix} 0 \\ 10 \\ -15 \end{pmatrix}$ and $\overrightarrow{PR} = \begin{pmatrix} -15 \\ 10 \\ 0 \end{pmatrix}$.

(ii) Show that the vector $\begin{pmatrix} 2 \\ 3 \\ 2 \end{pmatrix}$ is perpendicular to the plane through P, Q and R.

Hence find the equation of this plane.

A hole is then drilled perpendicular to triangle PQR. The hole passes through the triangle at the point T which divides the line PS in the ratio 2 : 1, where S is the midpoint of QR.

(iii) Write down the co-ordinates of S, and show that the point T has co-ordinates $\left(-5, 16\frac{2}{3}, 25\right)$.

(iv) Write down the vector equation of the line of the drill hole.

Hence determine whether or not this line passes through C.

OCR MEI Structured Mathematics C4 4754(A) June 2005 Q8

3 When a light ray passes from air to glass, it is deflected through an angle. The light ray ABC starts at point A(1, 2, 2), and enters a glass object at point B(0, 0, 2).

The surface of the glass object is a plane with normal vector **n**. The diagram shows a cross-section of the glass object in the plane of the light ray and **n**.

(i) Find the vector AB and a vector equation of the line AB.

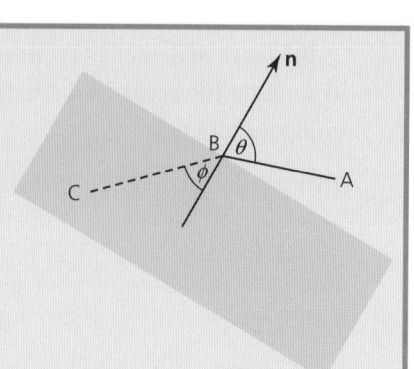

The surface of the glass object is a plane with equation $x + z = 2$. AB makes an acute angle θ with the normal to this plane.

(ii) Write down the normal vector **n**, and hence calculate θ, giving your answer in degrees.

The line BC has vector equation $\mathbf{r} = \begin{pmatrix} 0 \\ 0 \\ 2 \end{pmatrix} + \mu \begin{pmatrix} -2 \\ -2 \\ -1 \end{pmatrix}$. This line makes an acute angle ϕ with the normal to the plane.

(iii) Show that $\phi = 45°$.

(iv) Snell's Law states that $\sin \theta = k \sin \phi$, where k is a constant called the refractive index. Find k.

The light ray leaves the glass object through a plane with equation $x + z = -1$. Units are centimetres.

(v) Find the point of intersection of the line BC with the plane $x + z = -1$. Hence find the distance the light ray travels through the glass object.

OCR MEI Applications of Advanced Mathematics C4 4754/01A June 2009 Q7

4 The minimum distance from the point $P(1, a, 4)$ to the line $\mathbf{r} = \begin{pmatrix} -1 \\ a \\ 3 \end{pmatrix} + t \begin{pmatrix} 2 \\ -1 \\ -1 \end{pmatrix}$ is $\dfrac{\sqrt{14}}{2}$. Find the value(s) of the constant a.

Exam focus

1 The points P and Q have position vectors, relative to the origin O, given by

$\overrightarrow{OP} = -5\mathbf{i} - \mathbf{j} + 3\mathbf{k}$ and $\overrightarrow{OQ} = \mathbf{i} + 2\mathbf{j} + 4\mathbf{k}$.

The line l has vector equation

$\mathbf{r} = (1 + t)\mathbf{i} + (3 - 2t)\mathbf{j} + (5 + 2t)\mathbf{k}$.

(i) Show that l does not intersect the line passing through P and Q. [4]

(ii) Find the co-ordinates of the point A on l such that angle AQP is 90°. [4]

2 The straight line l has equation $\mathbf{r} = (2\mathbf{i} - \mathbf{j} - 3\mathbf{k}) + s(3\mathbf{i} - 5\mathbf{j} + 2\mathbf{k})$.

The plane p has equation $(\mathbf{r} - 15\mathbf{i}) \cdot (\mathbf{i} - 2\mathbf{j} + 2\mathbf{k}) = 0$.

The line l intersects the plane p at the point A.

(i) Find the position vector of A. [3]

(ii) Find the acute angle between l and p. [4]

(iii) Find a vector equation for the line which lies in p, passes through A and is perpendicular to l. [5]

3 With respect to the origin O, the points A, B and C have the following position vectors.

$$\overrightarrow{OA} = \begin{pmatrix} 1 \\ 2 \\ -3 \end{pmatrix}, \overrightarrow{OB} = \begin{pmatrix} 3 \\ -1 \\ 6 \end{pmatrix}, \overrightarrow{OC} = \begin{pmatrix} -2 \\ 4 \\ -1 \end{pmatrix}$$

The plane p is parallel to OB and contains A and C.

(i) Find the equation of p, giving your answer in the form $ax + by + cz = d$. [4]

(ii) Find the length of the perpendicular from B to the line through A and C. [3]

4 Two lines have equations $\mathbf{r} = \begin{pmatrix} 2 \\ k \\ 1 \end{pmatrix} + \lambda \begin{pmatrix} 3 \\ 2 \\ 2 \end{pmatrix}$ and $\mathbf{r} = \begin{pmatrix} 11 \\ -4 \\ -1 \end{pmatrix} + \mu \begin{pmatrix} -1 \\ 1 \\ 2 \end{pmatrix}$, where k is a constant. The lines intersect.

(i) Find the value of k and determine the co-ordinates of the point of intersection. **[4]**

(ii) Find the equation of the plane containing the two lines, giving your answer in the form $ax + by + cz = d$, where a, b, c and d are integers. **[5]**

11 Complex numbers

Working with complex numbers

EXERCISE 11.1

1 Simplify the following.

(i) i^3

(ii) $4i^4$

(iii) $-2i^6$

(iv) $5i^5$

(v) $(3i^3)^2$

(vi) $i^4 - i^3 + i^2 - i$

2 Simplify the following.

(i) $4 + 3i - 2 + 6i$

(ii) $4(2 - i) - 3(2 + i)$

(iii) $3i \times 4i$

(iv) $i(7 + i)$

(v) $(5 + 2i)(3 - i)$

(vi) $(2 - 3i)^2$

(vii) $(6 + 5i)(6 - 5i)$

(viii) $i(2 - i)(1 + i)$

(ix) $(1 + \sqrt{3}i)(1 + (\sqrt{3} - 2)i)$

3 Solve the following quadratic equations.

(i) $z^2 + 9 = 0$

(ii) $z^2 - 2x + 2 = 0$

(iii) $2z^2 - 6z + 5 = 0$ (iv) $z^2 + 2iz - 5 = 0$

4 The complex numbers z and w are given by $z = 5 - 2i$ and $w = 3 + 7i$.

Giving your answers in the form $x + iy$ and showing clearly how you obtain them, find the following.

(i) $4z - 3w$ (ii) z^*w

OCR MEI Further Pure Mathematics 1 4725 June 2009 Q3

(iii) z^*z (iv) $\text{Re}(z - z^*)$

(v) ww^* (vi) $iw + iz$

(vii) $w^* + w$ (viii) $\text{Im}[(w^*)^2]$

(ix) $(izw)^2$ (x) $[(z^* + w)^*]^2$

5 Show that for any complex number z:

(i) zz^* is a real number

(ii) $z + z^*$ is a real number.

Dividing and finding square roots of complex numbers

EXERCISE 11.2

1 Express these complex numbers in the form $x + iy$.

(i) $\dfrac{2}{i}$

(ii) $\left(\dfrac{15}{i}\right)^2$

(iii) $\dfrac{3}{1-i}$

(iv) $\dfrac{1}{2+3i}$

(v) $\dfrac{2i}{3-i}$

(vi) $\dfrac{5+3i}{4+3i}$

(vii) $\dfrac{(2-3i)^2}{3+i}$

(viii) $\dfrac{2+3i}{(2-i)^2}$

(ix) $\dfrac{4}{i} - \dfrac{3}{2-i}$

2 Find real numbers a and b that satisfy the following.

  (i) $(a + bi)^2 = -5 - 12i$

  (ii) $(a + bi)^2 = -24 + 10i$

3 Find complex numbers in the form $z = x + iy$ that satisfy the following.

  (i) $2z - z^* = 3 + 6i$

  (ii) $(1 - i)z = 2 + 8i$

  (iii) $iz + (2 + i)z^* = 10 - 2i$

  (iv) $(2 + i)z + (3 - 2i)z^* = 32$

4 Find all the complex numbers z that satisfy $iz^2 = 4z^*$.

5 Find the values of a and b such that $\dfrac{a}{1-i} + \dfrac{b}{1+i} = 1 + 2i$.

6 The complex number u is defined by $u = \dfrac{4i}{a-3i}$.

Express u in the form $x + iy$, where x and y are real.

Representing complex numbers geometrically and finding the modulus

EXERCISE 11.3

1 Represent each of these complex numbers on the Argand diagram.

(i) $4i$

(ii) $1 - i$

(iii) $-5 + 2i$

(iv) $3 + 4i$

(v) -4

(vi) $-4 - 3i$

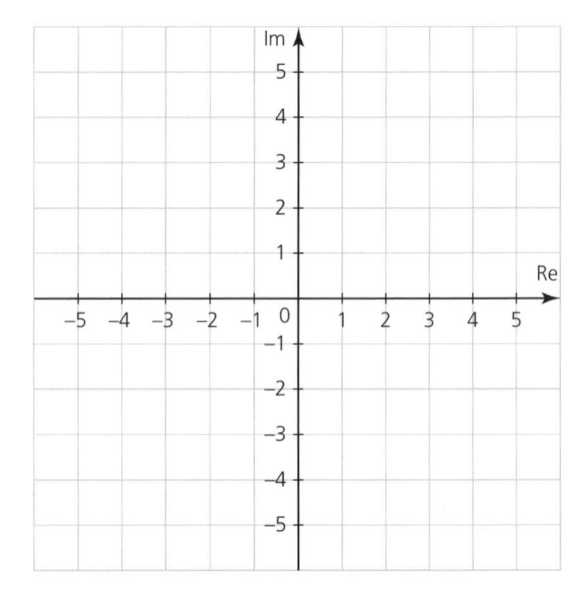

2 Given that $z = 1 + 2i$, plot the following points on the Argand diagram.

(i) z

(ii) $z + 2$

(iii) $z - 3i$

(iv) z^*

(v) iz

(vi) iz^2

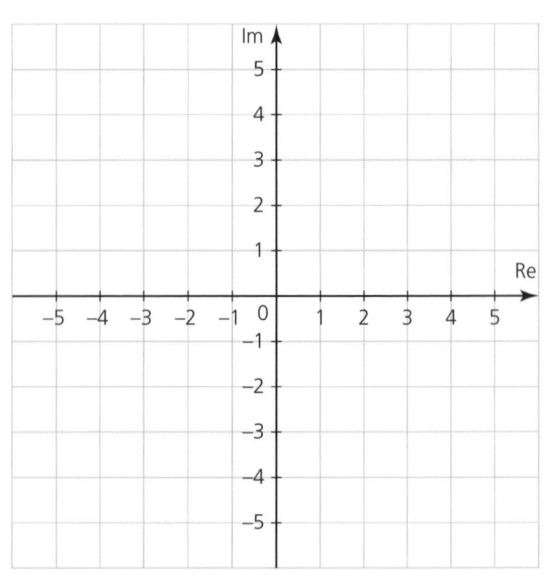

3 Given that $z = 2 - i$ and $w = -3 + 2i$, represent the following complex numbers on the Argand diagram.

(i) z

(ii) w

(iii) $z + w$

(iv) $w - z$

(v) zw

(vi) $z^* - w$

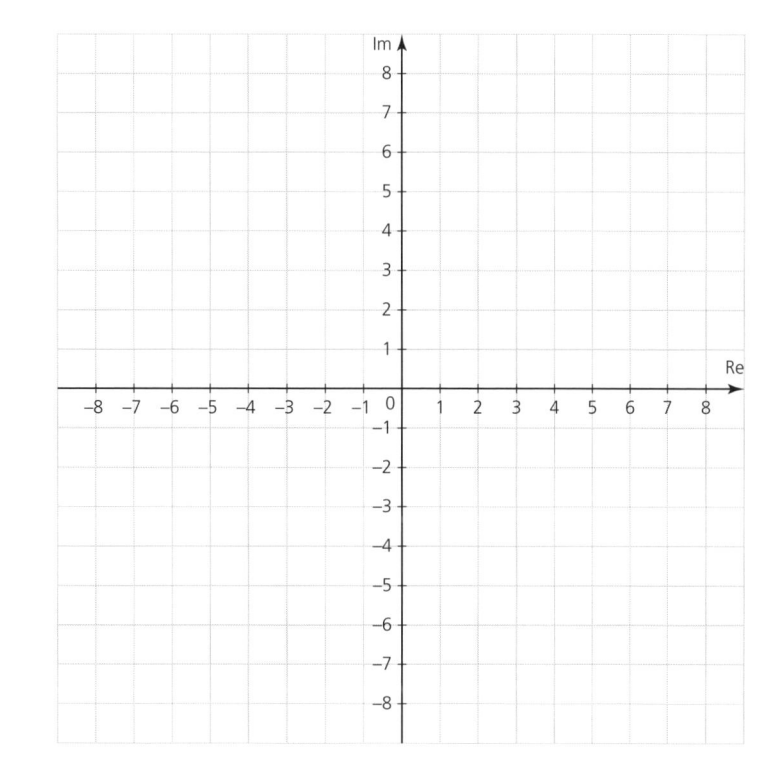

4 Find the modulus of the following complex numbers.

(i) $\sqrt{3} - i$

(ii) $5 + 12i$

(iii) $4i$

(iv) $-\sqrt{2} + \sqrt{5}i$

5 Given that $u = 1 + i$, $v = -4 - 3i$, $w = 2i$, find the following.

(i) $|u|$

(ii) $|v|$

(iii) $|u + v|$

(iv) $|u - w|$

(v) $|w^*|$

(vi) $|u^* + v - w|$

(vii) $\left|\dfrac{u}{w^*}\right|$

(viii) $|uv|$

6 Simplify.

(i) $|3 - 4i|$

(ii) $|x - 1 + i|$

(iii) $|(x+1) + i(y+1)|$

(iv) $|\cos\theta + i\sin\theta|$

Loci with complex numbers

EXERCISE 11.4

1 Sketch on each Argand diagram the set of points z represented by the equation or inequality.

(i) $|z| = 3$

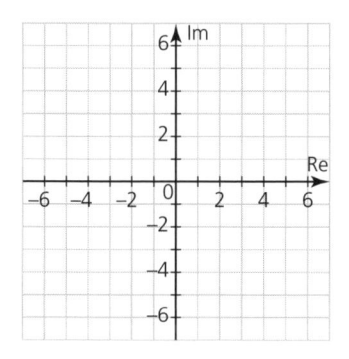

(ii) $|z - 2| = 2$

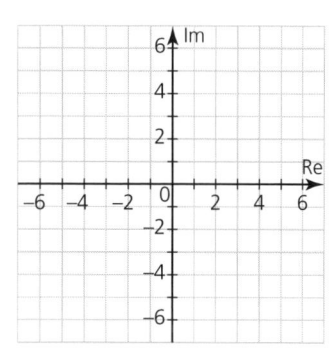

(iii) $|z + i| \leq 3$

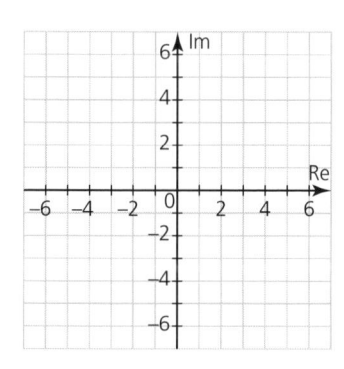

(iv) $|z - (2 + i)| \leq 2$

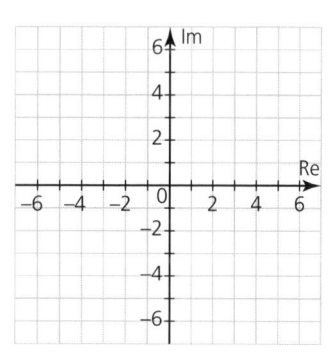

(v) $|z + 1 - i| \leq 2$

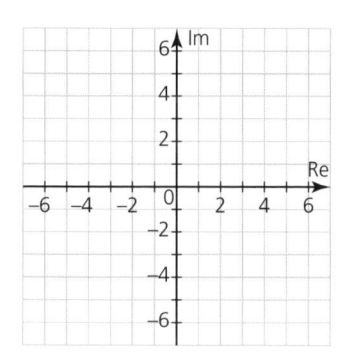

(vi) $|z + 2 + 2i| \geq 2$

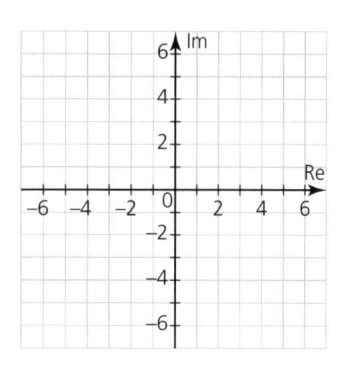

2 Sketch on each Argand diagram the set of points z represented by the equation or inequality.

(i) $|z| = |z - 2|$

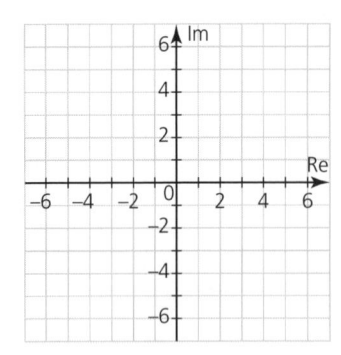

(ii) $|z - 2i| = |z + 2|$

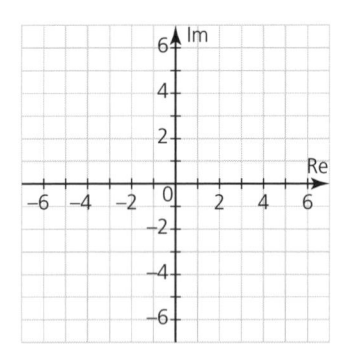

(iii) $|z + i| \leq |z + 3 - i|$

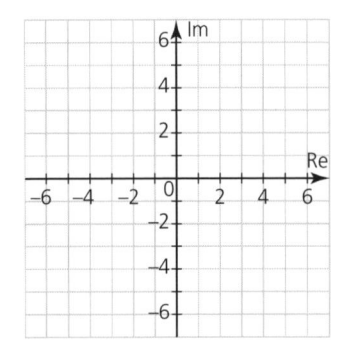

(iv) $|z - 2 - i| \geq |z + 4 + i|$

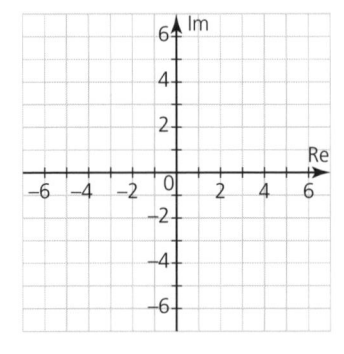

(v) $|z + 2 + 3i| \leq |z - 3 - 2i|$

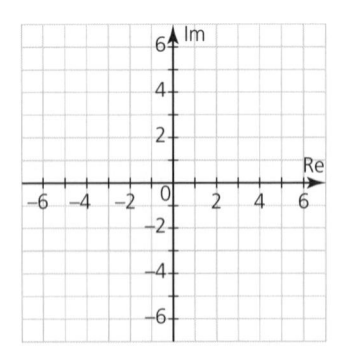

(vi) $|z + 4 + i| \leq |z + 3i|$

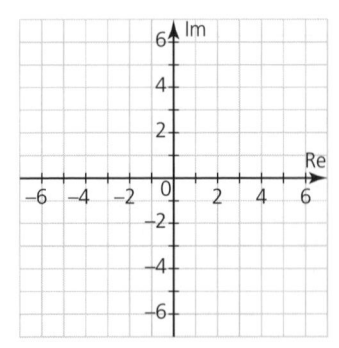

3 Explain what the difference would be in the loci sketched in questions **1** and **2** if the inequality signs used had been > or < rather than ⩾ or ⩽.

4 The complex number w is defined by $w = 1 - 2i$.

(i) Showing your working, express w^2 in the form $x + iy$, where x and y are real.

Find the modulus of w^2.

(ii) Shade on the Argand diagram the region whose points represent the complex numbers z which satisfy $|z - w^2| \leq |w^2|$.

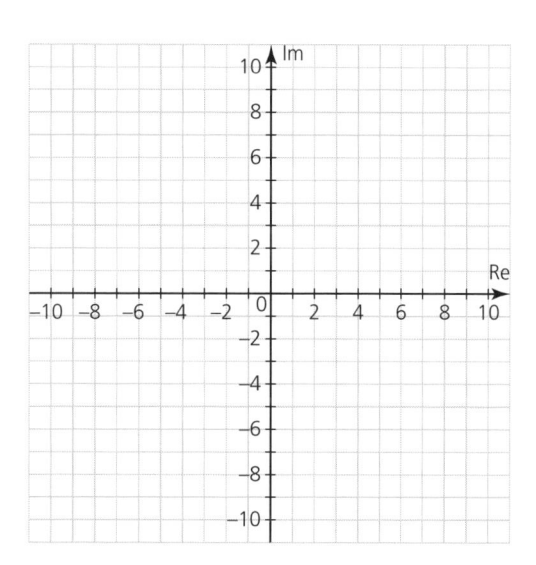

5 On the Argand diagram sketch the loci z and find in each case the greatest and least values of $|z|$.

(i) $|z - 3 - 2i| = 2$

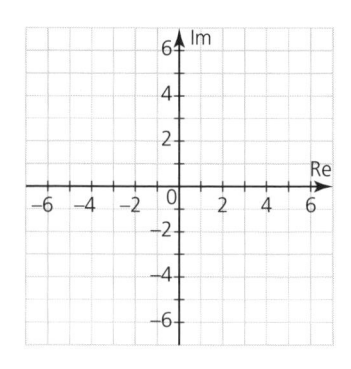

(ii) $|z + 1 - i| = 3$

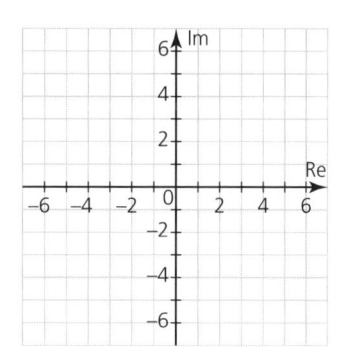

6 On the Argand diagram sketch the locus of points z such that $|z| = |z - 2 - 2i|$ and find the minimum value of $|z|$ for this locus.

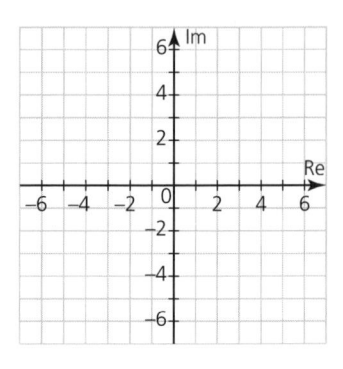

7 Sketch the locus of points z represented by the following.

(i) $\arg(z) = \dfrac{\pi}{3}$

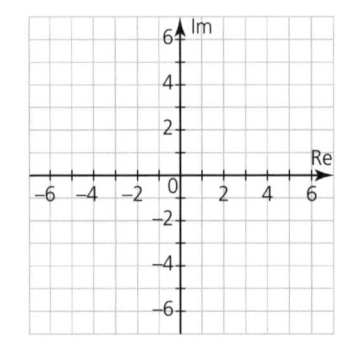

(ii) $\arg(z-1) = \dfrac{\pi}{2}$

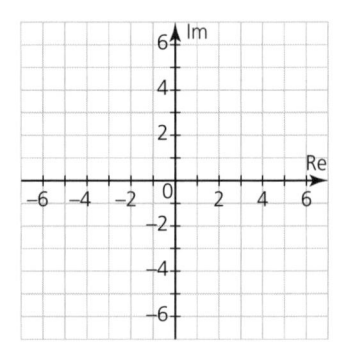

(iii) $\arg(z - 3i) = -\dfrac{3\pi}{4}$

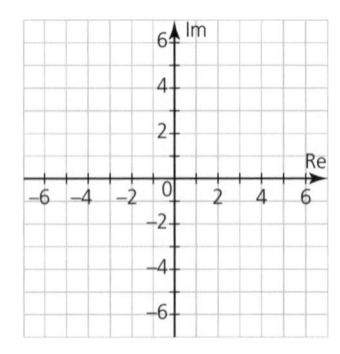

(iv) $\arg(z - 1 + 2i) = \pi$

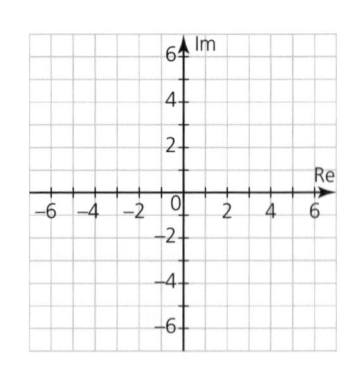

(v) $\arg(z+i) \geq \dfrac{\pi}{4}$

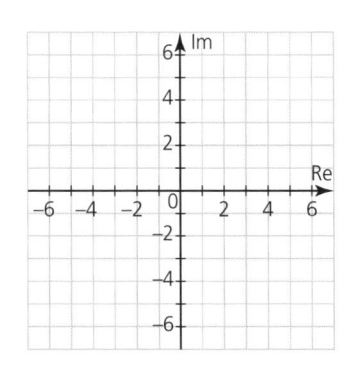

(vi) $-\dfrac{\pi}{6} \leq \arg(z-2) \leq \dfrac{\pi}{2}$

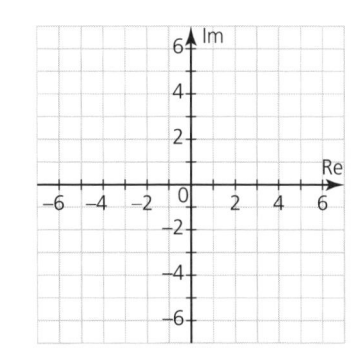

8 Write down the equation of the locus represented in the Argand diagram shown.

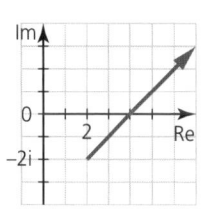

OCR MEI Further Concepts for Advanced Mathematics FP1 4755 January 2009 Q4

9 (i) Write down the equation of the locus represented by the perimeter of the circle in the Argand diagram.

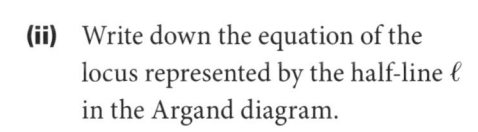

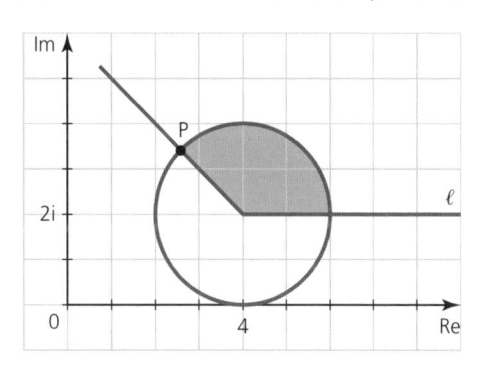

(ii) Write down the equation of the locus represented by the half-line ℓ in the Argand diagram.

(iii) Express the complex number represented by the point P in the form $a + bi$, giving the exact values of a and b.

(iv) Use inequalities to describe the set of points that fall within the shaded region (excluding its boundaries) in the Argand diagram

OCR MEI Further Concepts for Advanced Mathematics FP1 4755 June 2009 Q8

10 (i) On the Argand diagram, show the region where $-\dfrac{\pi}{2} < \arg(z-2-i) < \dfrac{\pi}{4}$.

(ii) Determine whether the point $43 + 47i$ lies within this region.

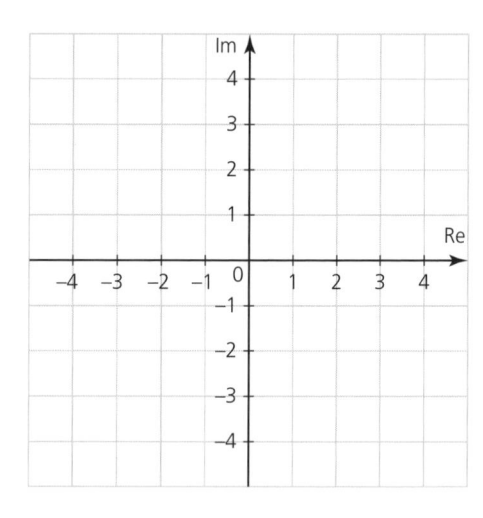

OCR MEI Further Concepts for Advanced Mathematics FP1 4755 January 2010 Q8(b)

11 (i) Sketch the locus $|z-3-4i| = 2$ on the Argand diagram.

(ii) On the same diagram, sketch the locus $\arg(z-4) = \dfrac{1}{2}$.

(iii) Indicate clearly on your sketch the set of points that satisfy **both** $|z-3-4i| \leq 2$ and $\arg(z-4) \geq \dfrac{1}{2}$.

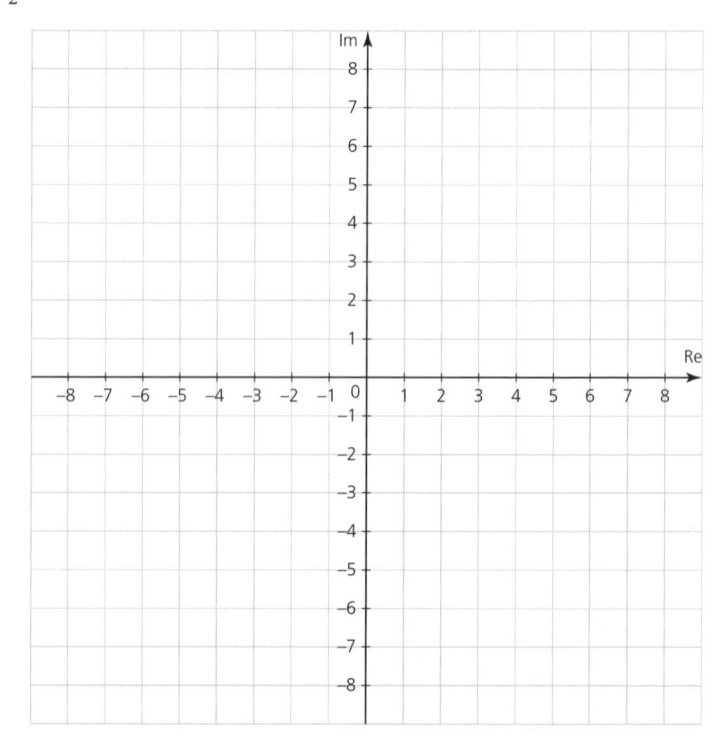

OCR MEI Further Concepts for Advanced Mathematics FP1 4755 June 2005 Q5

12 (i) Sketch on an Argand diagram the locus, C, of points for which $|z-4|=3$.

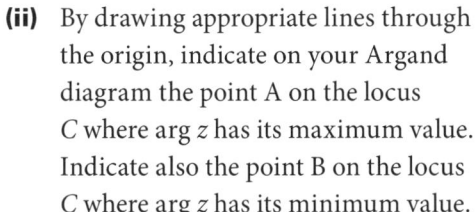

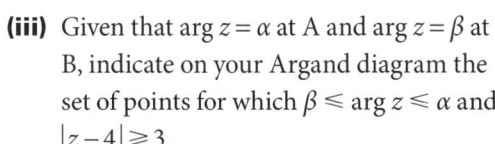

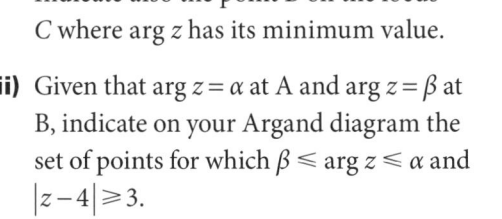

(ii) By drawing appropriate lines through the origin, indicate on your Argand diagram the point A on the locus C where arg z has its maximum value. Indicate also the point B on the locus C where arg z has its minimum value.

(iii) Given that arg $z = \alpha$ at A and arg $z = \beta$ at B, indicate on your Argand diagram the set of points for which $\beta \leq \arg z \leq \alpha$ and $|z-4| \geq 3$.

(iv) Calculate the value of α and the value of β.

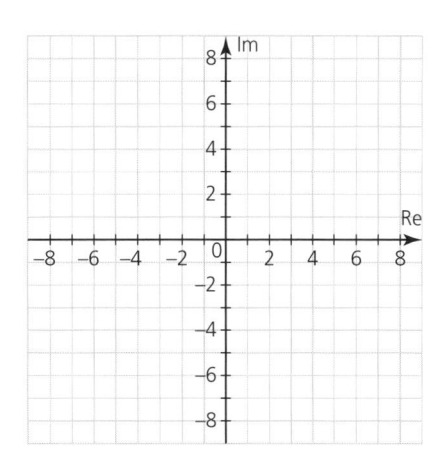

OCR MEI Further Concepts for Advanced Mathematics FP1 4755 January 2012 Q8

13 (i) Shade the region in the complex plane satisfied by

$|z - 2i| \leq 2$ and $\arg(z - 2i) \geq \dfrac{\pi}{4}$ and $\text{Im}(z) \geq 3$

where $\text{Im}(z)$ denotes the imaginary part of z.

(ii) Calculate the greatest possible value of Re z for points lying in the shaded region, where Re z denotes the real part of z.

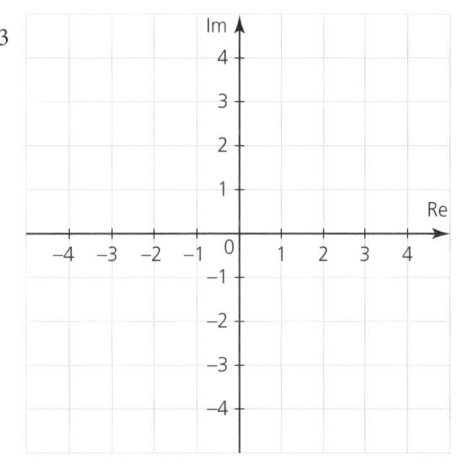

14 The complex number $3 - 3i$ is denoted by a.

(i) Find $|a|$ and arg a.

(ii) Sketch on the Argand diagram the loci given by the following.

(a) $|z-a| = 3\sqrt{2}$

(b) $\arg(z-a) = \dfrac{\pi}{4}$

(c) $|z-a| = |z-6|$

(iii) Indicate, by shading, the region of the Argand diagram for which $|z-a| \leqslant 3\sqrt{2}$ and $0 \leqslant \arg(z-a) \leqslant \frac{\pi}{4}$ and $|z-a| \geqslant |z-6|$.

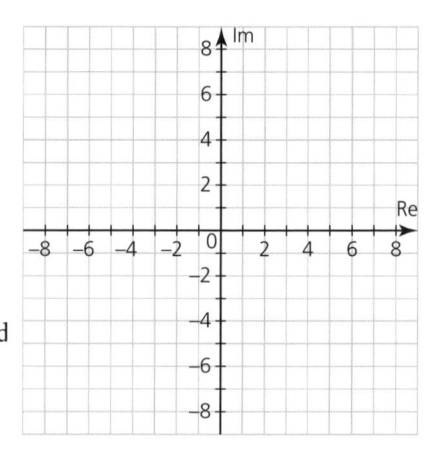

OCR MEI Further Pure Mathematics 1 4725 June 2009 Q6

15 The complex number $w = 3 - i$.

(i) Sketch the locus of point z such that $|z-w| = 2$.

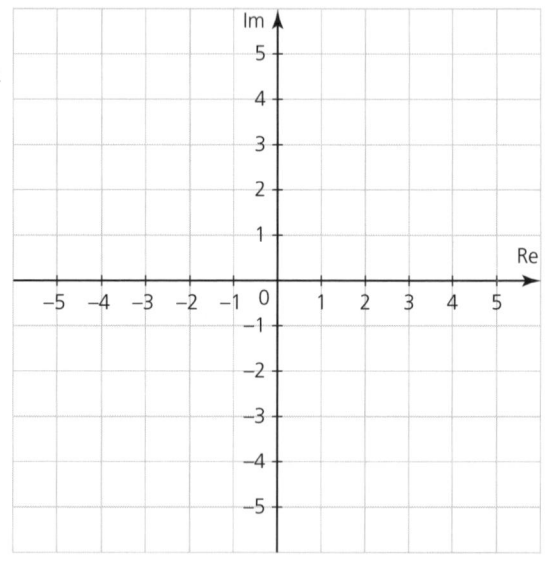

(ii) Given that $|z-w| = 2$, find the maximum and minimum value of the following.

(a) $|z|$ (b) $|z-3|$ (c) $\arg(z)$

Modulus–argument form and exponential form

EXERCISE 11.5

1 Write these complex numbers in modulus–argument (or polar) form, $z = r(\cos\theta + i\sin\theta)$, where $r > 0$ and $-\pi \leq \theta \leq \pi$.

(i) $-2i$

(ii) $4 - 3i$

(iii) $-5 + 12i$

(iv) $1 + \sqrt{3}i$

(v) 8

(vi) $-1 - i$

2 Write these complex numbers in the form $a + bi$.

(i) $2(\cos 45° + i\sin 45°)$

(ii) $\cos\left(-\frac{\pi}{3}\right) + i\sin\left(-\frac{\pi}{3}\right)$

(iii) $12\left(\cos\frac{2\pi}{3} + i\sin\frac{2\pi}{3}\right)$

(iv) $0.5(\cos(-90°) + i\sin(-90°))$

(v) $6\left(\cos\left(\frac{3\pi}{8}\right) + i\sin\left(\frac{3\pi}{8}\right)\right)$

(vi) $\sqrt{2}(\cos 135° + i\sin 135°)$

Note that $z = r(\cos\theta + i\sin\theta)$ can be written in abbreviated form as $z = r\operatorname{cis}\theta$.

3 The complex numbers u, v and w are given by
$u = 4\operatorname{cis}30°$, $v = 2\operatorname{cis}60°$, $w = \operatorname{cis}(-90°)$.
Find the following, in modulus–argument form, $r\operatorname{cis}\theta$.

(i) u^* (ii) $u \times v$ (iii) $v \times w$

(iv) $u \times w$ (v) $u \times w^*$ (vi) $\dfrac{v}{u}$

(vii) $\dfrac{u}{v^*}$ (viii) $\dfrac{u^2}{v^3}$ (ix) $\dfrac{1}{v}$

(x) iu (xi) w^4 (xii) $\dfrac{iv}{u}$

4 For any complex numbers z_1 and z_2 decide if the following statements are true or false.

(i) $|z_1 z_2| = |z_1| \times |z_2|$ (ii) $|z_1 + z_2| = |z_1| + |z_2|$

(iii) $|z_1|^2 = |z_1^2|$ (iv) $\dfrac{|z_1|}{|z_2|} = \left|\dfrac{z_1}{z_2}\right|$

(v) $|z_1^*| = |z_1|$ (vi) $\arg\dfrac{z_1}{z_2} = \arg z_1 - \arg z_2$

5 Two complex numbers are $\alpha = -\sqrt{3} - i$ and $\beta = 4\operatorname{cis}\frac{1}{2}$.

Find the modulus and argument of each of the following complex numbers, giving the argument in radians between $-\pi$ and π. Illustrate these 4 complex numbers on an Argand diagram.

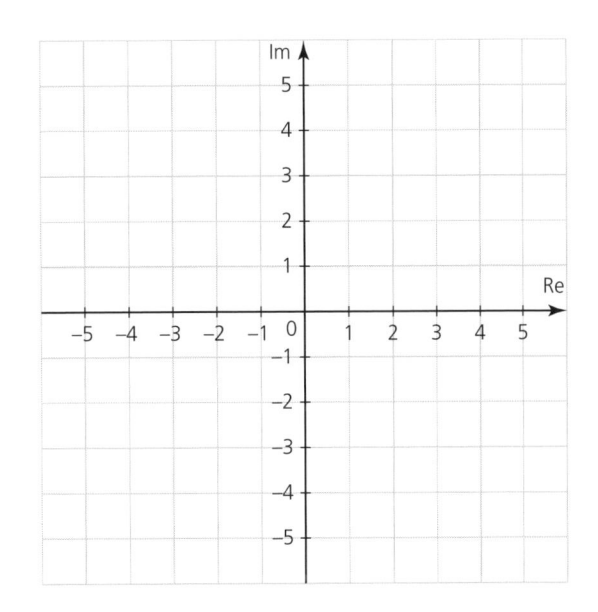

(i) α

(ii) β

(iii) $\alpha\beta$

(iv) $\dfrac{\alpha}{\beta}$

6 It is given that $m = -4 + 2i$.

  (i) Express $\frac{1}{m}$ in the form $a + bi$.

  (ii) Express m in modulus–argument form.

  (iii) Represent the following loci on the Argand diagram.

  (a) $|z - m| = 3$

  (b) $\arg(z - m) = \frac{\pi}{4}$

  (c) $|z - m| = |z - 2i|$

  (iv) Hence shade the region that satisfies

  $|z - m| \leq 3$ and

  $0 < \arg(z - m) < \frac{\pi}{4}$ and

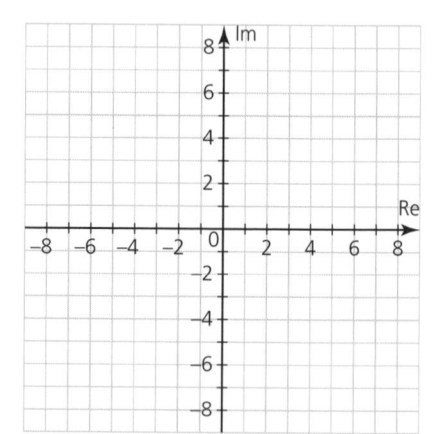

  $|z - m| \geq |z - 2i|$.

OCR MEI Further Concepts for Advanced Mathematics FP1 4755/01 January 2007 Q8

7 Show that for any non-zero complex number z, $\dfrac{1}{z} = \dfrac{z^*}{|z|^2}$.

8 This question introduces an alternative method for finding the square root of a complex number.

The complex number z is such that $z^2 = 5 - 12i$.

(i) Write z^2 in modulus–argument form $r\operatorname{cis}\theta$.

(ii) Take the square root of both sides of the equation $z^2 = r\operatorname{cis}\theta$ giving $z = (r\operatorname{cis}\theta)^{\frac{1}{2}}$.

(iii) Find one possible value of z using that fact that $(r\operatorname{cis}\theta)^{\frac{1}{2}} = r^{\frac{1}{2}}\operatorname{cis}\left(\tfrac{1}{2}\theta\right) = \sqrt{r}\operatorname{cis}\left(\tfrac{1}{2}\theta\right)$.

(iv) Find the other value of z using the fact that the other complex number satisfying $z^2 = 5 - 12i$ can be found by rotating the first value by 180° around the Argand diagram.

(v) Convert both values for z to the form $x + iy$.

9 Use the method described in question **8** to find $\sqrt{21 - 20i}$.

[Hint: start with $z^2 = 21 - 20i$.]

10 The complex number u is defined by $u = \dfrac{3}{2+ai}$, where the constant a is real.

  (i) Express u in the form $x + iy$, where x and y are real.

  (ii) Find the value of the constant a such that:

  (a) $\arg u^* = \dfrac{\pi}{4}$

  (b) $|u| = \sqrt{2}$.

11 Convert these complex numbers to exponential form, $z = re^{i\theta}$.

  (i) $2\operatorname{cis}\dfrac{\pi}{2}$ **(ii)** $-1 + i$

  (iii) $2 - 3i$ **(iv)** $5i$

12 Write these complex numbers in the form $x + iy$. Plot each one on the Argand diagram.

(i) $e^{\frac{-i}{3}}$

(ii) $2e^{-i}$

(iii) $4e^{i\pi}$

(iv) $3e^{-\frac{-i}{6}}$

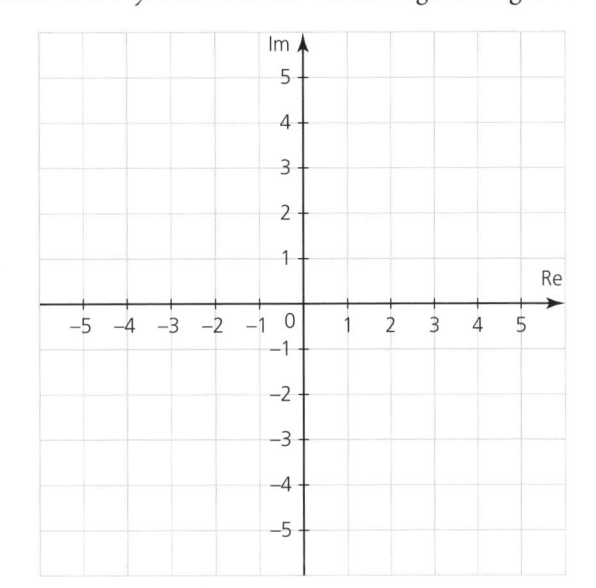

13 (i) Find $4e^{\frac{-i}{3}} \times 3\text{cis}\left(\frac{-}{6}\right)$ in exponential form.

(ii) Find $-2i + 2\text{cis}\left(\frac{-}{6}\right)$ in:

(a) the form $x + iy$

(b) polar form.

Complex numbers and equations

EXERCISE 11.6

1. The cubic equation $2z^3 - z^2 + 4z + k = 0$, where k is real, has a root $z = 1 + 2i$. Write down the other complex root. Hence find the real root and the value of k.

OCR MEI Further Concepts for Advanced Mathematics FP1 4755 June 2010 Q3

2. Given that $z = 6$ is a root of the cubic equation $z^3 - 10z^2 + 37z + p = 0$, find the value of p and the other roots.

OCR MEI Further Concepts for Advanced Mathematics FP1 4755 January 2012 Q3

3 (i) Find the solutions to the equation $iz^2 - z + 2i = 0$.

(ii) The roots of the equation are NOT a conjugate pair. Why not?

4 (i) Find the solutions of $z^4 - 16 = 0$.

(ii) Plot the solutions on an Argand diagram. What do you notice?

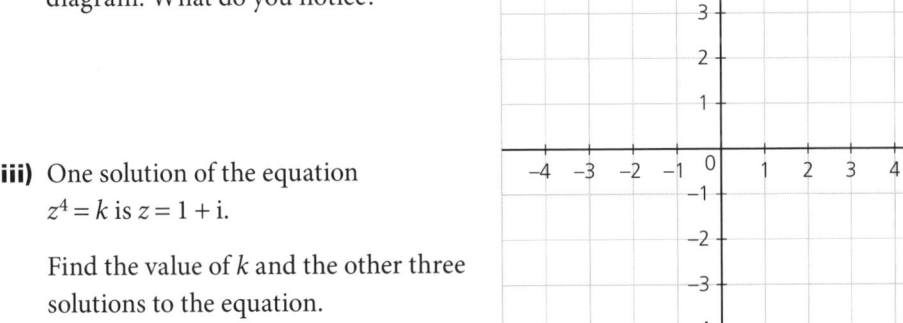

(iii) One solution of the equation $z^4 = k$ is $z = 1 + i$.

Find the value of k and the other three solutions to the equation.

5 The equation $x^4 + Ax^3 + Bx^2 + Cx + D = 0$, where A, B, C, and D are real numbers, has roots $2 + i$ and $-2i$.

(i) Write down the other roots of the equation.

(ii) Find the values of A, B, C, and D.

OCR MEI Further Concepts for Advanced Mathematics FP1 4755 June 2005 Q9

6 You are given that the complex number $\alpha = 1 + i$ satisfies the equation $z^3 + 3z^2 + pz + q = 0$, where p and q are real constants.

(i) Find α^2 and α^3 in the form $a + bi$. Hence show that $p = -8$ and $q = 10$.

(ii) Find the other two roots of the equation.

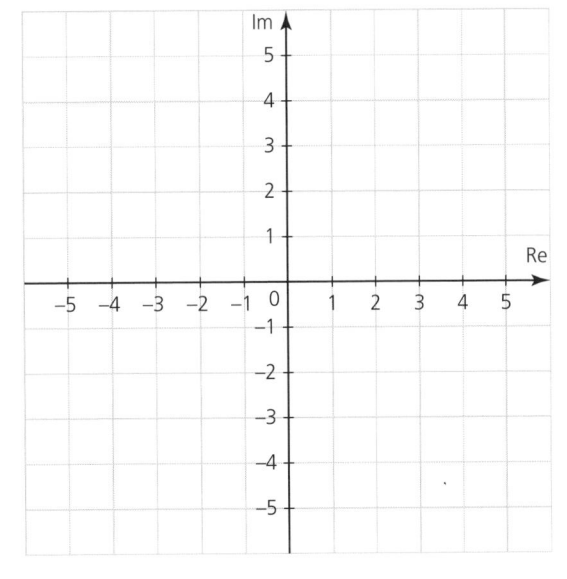

(iii) Represent the three roots on an Argand diagram.

OCR MEI Further Concepts for Advanced Mathematics FP1 4755 January 2006 Q8

7 The cubic equation $x^3 + Ax^2 + Bx + 15 = 0$, where A and B are real numbers, has a root $x = 1 + 2i$.

(i) Write down the other complex root.

(ii) Explain why the equation must have a real root.

(iii) Find the value of the real root and the values of A and B.

8 (i) Show that $z = 3$ is a root of the cubic equation $z^3 + z^2 - 7z - 15 = 0$ and find the other roots.

(ii) Show the roots on the Argand diagram.

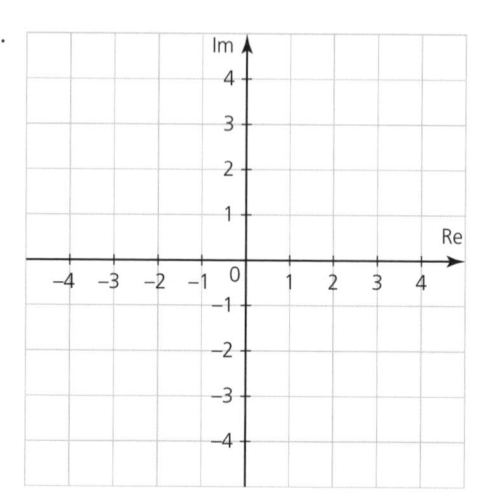

OCR MEI Further Concepts for Advanced Mathematics FP1 4755/01 January 2008 Q3

9 Two complex numbers, α and β, are given by $\alpha = 1 + i$ and $\beta = 2 - i$.

  (i) Express $\alpha + \beta$, $\alpha\alpha^*$ and $\alpha\beta$ in the form $a + bi$.

  (ii) Find a quadratic equation with roots α and α^*.

(iii) α and β are roots of a quartic equation with real coefficients.
Write down the two other roots and find this quartic equation in the form
$z^4 + Az^3 + Bz^2 + Cz + D = 0$.

OCR MEI Further Concepts for Advanced Mathematics FP1 4755 January 2009 Q9

10 The complex number z satisfies $z^n = a + b\text{i}$. When plotted on an Argand diagram, two adjacent roots are given by $\operatorname{cis}\dfrac{\pi}{24}$ and $\operatorname{cis}\dfrac{13}{24}$. Find the value of n and the exact values of a and b.

Stretch and challenge

1. Solve the equation $e^{x+iy} = 3$, where x and y are real.

2. The complex number z is defined by $z = \cos\theta + i\sin\theta$.

   (i) Show that $z^n + \dfrac{1}{z^n} = 2\cos n\theta$ and $z^n - \dfrac{1}{z^n} = 2i\sin n\theta$.

   (ii) Expand $\left(z + \dfrac{1}{z}\right)^4$ using the binomial theorem.
   Hence show that $\cos^4\theta = \dfrac{1}{8}(\cos 4\theta + 4\cos 2\theta + 3)$.

(iii) (a) Show that $\tan\theta = \dfrac{z - z^{-1}}{\mathrm{i}(z + z^{-1})}$.

(b) Use this result to prove that $\cos 2\theta = \dfrac{1 - \tan^2\theta}{1 + \tan^2\theta}$.

3 De Moivre's theorem states that $(r\operatorname{cis}\theta)^n = r^n \operatorname{cis}(n\theta)$.

(i) Use De Moivre's theorem to find w^{12} given that $w = \sqrt{3} - \mathrm{i}$.

(ii) By writing $\frac{1}{32}i$ in polar form, use De Moivre's theorem to find the solutions to the equation $z^5 = \frac{1}{32}i$.

(iii) (a) Expand $(\cos\theta + i\sin\theta)^5$ using the binomial theorem.

(b) By equating imaginary parts and using De Moivre's theorem, show that $\sin 5\theta = 16\sin^5\theta - 20\sin^3\theta + 5\sin\theta$.

(c) Hence solve the equation $16x^5 - 20x^3 + 5x - 1 = 0$, giving your answers in exact trigonometric form.

4 (i) The **gamma function** Γ extends the factorial function to all of the real numbers except for the negative integers and zero.

It takes values of the factorial function at positive integers: $\Gamma(n+1) = n!$ and has the property that for any values, $\Gamma(x) = (x-1)\Gamma(x-1)$.

A formula useful for finding other values of the gamma function is Euler's reflection formula: $\Gamma(z)\Gamma(1-z) = \dfrac{\pi}{\sin(\pi z)}$.

Find the **exact** value of $\Gamma\left(\frac{1}{2}\right)$, and hence the exact value of $\Gamma\left(\frac{5}{2}\right)$.

(ii) Consider the equation $x^6 + (2-k)x^4 + (25-2k)x^2 - 25k = 0$, where k is a **real** constant.

Given that $c = \sqrt{2} + \sqrt{3}i$ is a root of this equation, find the other five roots.

(iii) The vertices of a regular octagon, shown on the Argand diagram, represent the roots of a complex degree 8 polynomial. One root is shown at the point A, 1 + 0i.

Write the polynomial in the form
$p(z) = (z - (a + bi))^n + q$.

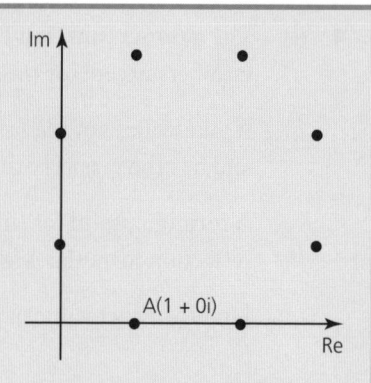

NZQA Scholarship Calculus 2011 Q5

5 (i) If $z = \sqrt{\frac{1}{2}\left(a + \sqrt{a^2 + b^2}\right)} + i\sqrt{\frac{1}{2}\left(-a + \sqrt{a^2 + b^2}\right)}$ is a complex number, with $i^2 = -1$ and a, b real numbers, find z^2 in the form $p + iq$.

The Mandelbrot set is constructed by plotting in black all complex numbers, $c = a + ib$, such that:

if $f(z) = z^2 + c$ then $|f^n(0)| < 2$ for all $n \in \{1, 2, 3, ...\}$

where $|f^n(z)|$ represents the modulus of the complex number $\underbrace{f(f(f(f(...f(z)))))}_{n}$.

That is, $f^2(z) = f(f(z)) = (z^2 + c)^2 + c$. The sequence starts with $z = 0$, so $f(0) = c$, $f^2(0) = c^2 + c$, etc.

(ii) Use the definition of the Mandelbrot set above for $f(z) = z^2 + c$ beginning with $z = 0$, so $f(0) = c$.

(a) Show that $c = 1 + 2i$ is not part of the Mandelbrot set, but $c = i$ is.

(b) For $f(z) = z^2 + c$, with $c = \sqrt{\frac{1}{2}\left(a + \sqrt{a^2 + b^2}\right)} + i\sqrt{\frac{1}{2}\left(-a + \sqrt{a^2 + b^2}\right)}$, and $b^2 = 3a^2$, find $f^2(0)$ in terms of a.

Hence show that when $a = \frac{1}{8}$, $\left|f^2(0)\right| = \frac{1}{4}\sqrt{(5 + 2\sqrt{3})}$.

NZQA Scholarship Calculus 2007 Q4

Exam focus

1 The complex numbers $1 - 2i$ and $3 - i$ are denoted by z and w respectively.

(i) Showing your working express each of the following in the form $x + iy$. [8]

    (a) $z - 3w$ (b) zw

    (c) $\dfrac{1}{w}$ (d) $\dfrac{w}{z^*}$

(ii) The complex number u is given by $u = \dfrac{z^2}{w}$.

    (a) Express u in the form $x + iy$, where x and y are real. [4]

    (b) Sketch on the Argand diagram the locus of the complex number z such that $|z - u| = |u|$. [3]

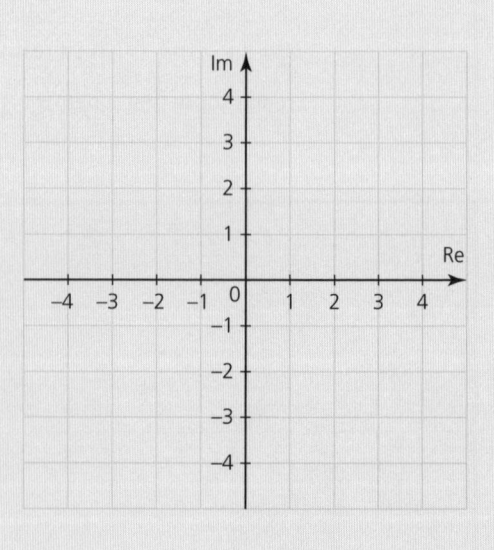

284

2 The complex numbers u and w satisfy the equations $u - w = 6i$ and $uw = 13$.

Solve the equations for u and w, giving all answers in the form $x + iy$, where x and y are real. [5]

3 (i) On a single Argand diagram, sketch the locus of each set of points.

   (a) $|z - 3i| = 2$ [2]

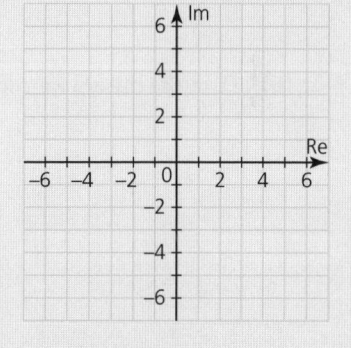

   (b) $\arg(z + 1) = \tfrac{1}{4}\pi$ [2]

(ii) Indicate clearly on your Argand diagram the set of points for which

$|z - 3i| \leq 2$ and $\arg(z + 1) \leq \tfrac{1}{4}\pi$. [2]

(iii) (a) By drawing an appropriate line through the origin, indicate on your Argand diagram the point for which $|z - 3i| = 2$ and $\arg z$ has its minimum possible value. [2]

   (b) Calculate the value of $\arg z$ at this point. [2]

OCR MEI Further Concepts for Advanced Mathematics FP1 4755/01 January 2008 Q8

4 The complex number $1 + 2i$ is denoted by u. The polynomial $x^4 - 3x^3 + 5x^2 - x - 10$ is denoted by $p(x)$.

  (i) Showing your working, verify that u is a root of the equation $p(x) = 0$, and write down a second complex root of the equation. [4]

  (ii) Find the other two roots of the equation $p(x) = 0$. [6]

5 (i) Without using a calculator, solve the equation $iw^2 = (-3 + 3i)^2$. [3]

(ii) (a) Sketch on the Argand diagram the region R consisting of points representing the complex numbers z where $|z - 3 + 3i| \leq 2$. [2]

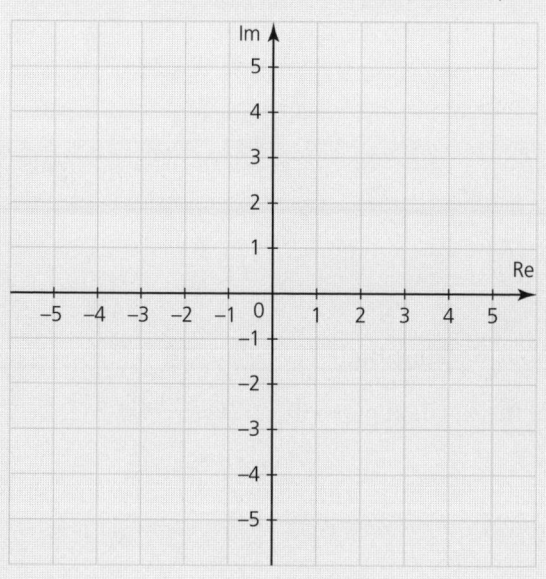

(b) For the complex numbers represented by points in the region R, it is given that $p \leq |z| \leq q$ and $\alpha \leq \arg z \leq \beta$.

Find the values of p, q, α and β. [6]

6 The polynomial p(z) is defined by $p(z) = z^3 - z^2 + kz - 27$, where k is a constant. It is given that $(z-3)$ is a factor of p(z).

(i) Find the value of k. [2]

(ii) Hence, showing all your working, find

(a) the three roots of the equation $p(z) = 0$ [5]

(b) the six roots of the equation $p(z^2) = 0$. [6]

PAST EXAMINATION QUESTIONS

1 Algebra

1 The polynomial $x^4 + 3x^2 + a$, where a is a constant, is denoted by p(x). It is given that $x^2 + x + 2$ is a factor of p(x). Find the value of a and the other quadratic factor of p(x). [4]

Cambridge International AS & A Level Mathematics, 9709/03 November 2007 Q2

2 Solve the inequality $|x-2| > 3|2x+1|$. [4]

Cambridge International AS & A Level Mathematics, 9709/03 May/June 2008 Q1

3 The polynomial $4x^3 - 4x^2 + 3x + a$, where a is a constant, is denoted by p(x). It is divisible by $2x^2 - 3x + 3$.

(i) Find the value of a. [3]

(ii) When a has this value, solve the inequality p(x) < 0, justifying your answer. [3]

Cambridge International AS & A Level Mathematics, 9709/03 November 2008 Q5

4 Solve the inequality $|x-3| > |2x|$. [4]

Cambridge International AS & A Level Mathematics, 9709/02 November 2008 Q1

5 Solve the inequality $2|x-3| > |3x+1|$. [4]

Cambridge International AS & A Level Mathematics, 9709/31 November 2010 Q1

6 The polynomial $x^3 + 4x^2 + ax + 2$, where a is a constant, is denoted by p(x). It is given that the remainder when p(x) is divided by $(x + 1)$ is equal to the remainder when p(x) is divided by $(x - 2)$.

(i) Find the value of a. [3]

(ii) When a has this value, show that $(x - 1)$ is a factor of p(x) and find the quotient when p(x) is divided by $(x - 1)$. [3]

Cambridge International AS & A Level Mathematics, 9709/23 November 2010 Q3

7 The cubic polynomial p(x) is defined by

p(x) = $6x^3 + ax^2 + bx + 10$,

where a and b are constants. It is given that $(x + 2)$ is a factor of p(x) and that, when p(x) is divided by $(x + 1)$, the remainder is 24.

(i) Find the values of a and b. [5]

(ii) When a and b have these values, factorise p(x) completely. [3]

Cambridge International AS & A Level Mathematics, 9709/22 May/June 2011 Q7

8 Solve the inequality $|x+2| > |\frac{1}{2}x - 2|$. [4]

Cambridge International AS & A Level Mathematics, 9709/22 November 2011 Q1

9 The polynomial $ax^3 - 3x^2 - 11x + b$, where a and b are constants, is denoted by p(x). It is given that ($x + 2$) is a factor of p(x), and that when p(x) is divided by ($x + 1$) the remainder is 12.

(i) Find the values of a and b. [5]

(ii) When a and b have these values, factorise p(x) completely. [3]

Cambridge International AS & A Level Mathematics, 9709/22 November 2011 Q7

2 Logarithms and exponentials

1 (i) Express 4^x in terms of y, where $y = 2^x$. [1]

(ii) Hence find the values of x that satisfy the equation

$$3(4^x) - 10(2^x) + 3 = 0,$$

giving your answers correct to 2 decimal places. [5]

Cambridge International AS & A Level Mathematics, 9709/02 November 2006 Q2

2 Solve the equation

$$\ln(x+2) = 2 + \ln x,$$

giving your answer correct to 3 decimal places. [3]

Cambridge International AS & A Level Mathematics, 9709/03 November 2008 Q1

3 Solve the equation $\ln(2 + e^{-x}) = 2$, giving your answer correct to 2 decimal places. [4]

Cambridge International AS & A Level Mathematics, 9709/3 May/June 2009 Q1

4 Solve the equation $3^{x+2} = 3^x + 3^2$, giving your answer correct to 3 significant figures. [4]

Cambridge International AS & A Level Mathematics, 9709/31 November 2009 Q2

5 (i) Given that $y = 2^x$, show that the equation

$2^x + 3^{(2-x)} = 4$

can be written in the form

$y^2 - 4y + 3 = 0$. [3]

(ii) Hence solve the equation

$2^x + 3(2^{-x}) = 4$,

giving the values of x correct to 3 significant figures where appropriate. [3]

Cambridge International AS & A Level Mathematics, 9709/21 May/June 2010 Q5

6 Solve the equation

$\ln(1 + x^2) = 1 + 2\ln x$,

giving your answer correct to 3 significant figures. [4]

Cambridge International AS & A Level Mathematics, 9709/31 November 2010 Q2

7 The variables x and y satisfy the equation $y = A(b^x)$, where A and b are constants. The graph of $\ln y$ against x is a straight line passing through the points $(1.4, 0.8)$ and $(2.2, 1.2)$, as shown in the diagram. Find the values of A and b, correct to 2 decimal places. [6]

Cambridge International AS & A Level Mathematics, 9709/23 November 2010 Q5

8 Use logarithms to solve the equation $3^x = 2^{x+2}$, giving your answer correct to 3 significant figures. [4]

Cambridge International AS & A Level Mathematics, 9709/22 May/June 2011 Q1

3 Trigonometry

1 (i) Prove the identity

$$\cos 4\theta + 4\cos 2\theta \equiv 8\cos^4\theta - 3.$$ [4]

(ii) Hence solve the equation

$$\cos 4\theta + 4\cos 2\theta = 2,$$

for $0° \leq \theta \leq 360°$. [4]

Cambridge International AS & A Level Mathematics, 9709/03 May/June 2005 Q6

2 (i) Prove the identity

$$\tan(x + 45°) - (\tan 45° - x) \equiv 2\tan 2x.$$ [4]

(ii) Hence solve the equation

$$\tan(x + 45°) - (\tan 45° - x) = 2,$$

for $0° \leq x \leq 180°$. [3]

Cambridge International AS & A Level Mathematics, 9709/02 November 2006 Q4

3 (i) Express $5\sin x + 12\cos x$ in the form $R\sin(x+\alpha)$, where $R > 0$ and $0° < \alpha < 90°$, giving the value of α correct to 2 decimal places. [3]

(ii) Hence solve the equation

$$5\sin 2\theta + 12\cos 2\theta = 11,$$

giving all solutions in the interval $0° < \theta < 180°$. [5]

Cambridge International AS & A Level Mathematics, 9709/03 November 2008 Q6

4 Solve the equation $\sec x = 4 - 2\tan^2 x$, giving all solutions in the interval $0° \leqslant x \leqslant 180°$. [6]

Cambridge International AS & A Level Mathematics, 9709/2 May/June 2009 Q5

5 (i) Express $3\cos x + 4\sin x$ in the form $R\cos(x-\alpha)$, where $R > 0$ and $0° < \alpha < 90°$, stating the exact valuae of R and giving the value of α correct to 2 decimal places. [3]

(ii) Hence solve the equation

$$3\cos x + 4\sin x = 4.5,$$

giving all solutions in the interval $0° < x < 360°$. [4]

Cambridge International AS & A Level Mathematics, 9709/22 November 2009 Q6

6 (i) Show that the equation $\tan(x + 45°) = 6\tan x$ can be written in the form $6\tan^2 x - 5\tan x + 1 = 0$. [3]

(ii) Hence solve the equation $\tan(x + 45°) = 6\tan x$, for $0° < x < 180°$. [3]

Cambridge International AS & A Level Mathematics, 9709/21 May/June 2010 Q3

7 Solve the equation

$$\sin\theta = 2\cos 2\theta + 1,$$

giving all solutions in the interval $0° \leqslant \theta \leqslant 360°$. [6]

Cambridge International AS & A Level Mathematics, 9709/31 May/June 2010 Q2

8 Solve the equation

$\cos(\theta + 60°) = 2\sin\theta$,

giving all solutions in the interval $0° \leqslant \theta \leqslant 360°$. [5]

Cambridge International AS & A Level Mathematics, 9709/31 November 2010 Q3

9 (i) Prove that $\sin^2 2\theta(\csc^2\theta - \sec^2\theta) \equiv 4\cos 2\theta$. [3]

(ii) Hence

(a) solve for $0° \leqslant x \leqslant 180°$ the equation $\sin^2 2\theta(\csc^2\theta - \sec^2\theta) = 3$, [4]

(b) find the exact value of $\csc^2 15° - \sec^2 15°$. [2]

Cambridge International AS & A Level Mathematics, 9709/22 May/June 2011 Q8

4 Differentiation

1 The equation of a curve is $y = x \sin 2x$, where x is in radians. Find the equation of the tangent to the curve at the point where $x = \frac{1}{4}\pi$. [4]

Cambridge International AS & A Level Mathematics, 9709/03 May/June 2007 Q3

2 The parametric equations of a curve are

$$x = a(2\theta - \sin 2\theta), \quad y = a(1 - \cos 2\theta).$$

Show that $\dfrac{dy}{dx} = \cot \theta$. [5]

Cambridge International AS & A Level Mathematics, 9709/03 November 2008 Q4

3 The diagram shows the curve $y = x^2\sqrt{(1 - x^2)}$ for $x \geq 0$ and its maximum point M.

Find the exact value of the x-coordinate of M. [4]

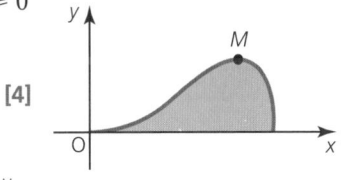

Cambridge International AS & A Level Mathematics, 9709/3 May/June 2009 Q10(i)

4 A curve has equation $y = e^{-3x} \tan x$. Find the x-coordinates of the stationary points on the curve in the interval $-\frac{1}{2}\pi < x < \frac{1}{2}\pi$. Give your answers correct to 3 decimal places. [6]

Cambridge International AS & A Level Mathematics, 9709/31 November 2009 Q4

5 The equation of a curve is

$x^2 y + y^2 = 6x.$

(i) Show that $\dfrac{dy}{dx} = \dfrac{6 - 2xy}{x^2 + 2y}$. [4]

(ii) Find the equation of the tangent to the curve at the point with coordinates (1, 2), giving your answer in the form $ax + by + c = 0$. [3]

Cambridge International AS & A Level Mathematics, 9709/21 May/June 2010 Q6

6 The diagram shows the curve $y = \sqrt{\left(\dfrac{1-x}{1+x}\right)}$.

(i) By first differentiating $\dfrac{1-x}{1+x}$, obtain an expression for $\dfrac{dy}{dx}$ in terms of x. Hence show that the gradient of the normal to the curve at the point (x, y) is $(1+x)\sqrt{(1-x^2)}$. [5]

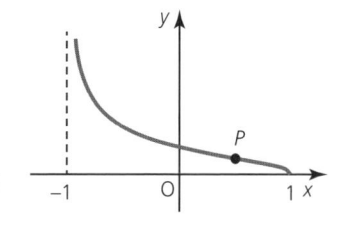

(ii) The gradient of the normal to the curve has its maximum value at the point P shown in the diagram. Find, by differentiation, the x-coordinate of P. [4]

Cambridge International AS & A Level Mathematics, 9709/31 May/June 2010 Q9

7 The diagram shows the curve $y = x^3 \ln x$ and its minimum point M.

Find the exact coordinates of M. [5]

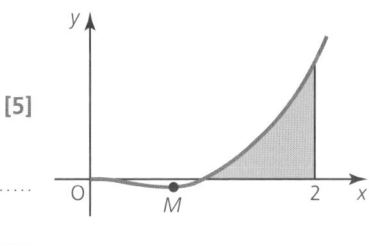

Cambridge International AS & A Level Mathematics, 9709/31 November 2010 Q9(i)

8 The equation of a curve is

$x^2 + 2xy - y^2 + 8 = 0$.

(i) Show that the tangent to the curve at the point $(-2, 2)$ is parallel to the x-axis. [4]

(ii) Find the equation of the tangent to the curve at the other point on the curve for which $x = -2$, giving your answer in the form $y = mx + c$. [5]

Cambridge International AS & A Level Mathematics, 9709/23 November 2010 Q8

9 The parametric equations of a curve are

$x = \ln(\tan t), \quad y = \sin^2 t$,

where $0 < t < \tfrac{1}{2}\pi$.

(i) Express $\dfrac{dy}{dx}$ in terms of t. [4]

(ii) Find the equation of the tangent to the curve at the point where $x = 0$. [3]

Cambridge International AS & A Level Mathematics, 9709/32 May/June 2011 Q5

10 A curve has equation $x^2 + 2y^2 + 5x + 6y = 10$. Find the equation of the tangent to the curve at the point $(2, -1)$. Give your answer in the form $ax + by + c = 0$, where a, b and c are integers. [6]

Cambridge International AS & A Level Mathematics, 9709/22 May/June 2011 Q5

11 The parametric equations of a curve are

$x = 1 + 2\sin^2\theta, \quad y = 4\tan\theta.$

(i) Show that $\dfrac{dy}{dx} = \dfrac{1}{\sin\theta \cos^3\theta}$. [3]

(ii) Find the equation of the tangent to the curve at the point where $\theta = \frac{1}{4}\pi$, giving your answer in the form $y = mx + c$. [4]

Cambridge International AS & A Level Mathematics, 9709/22 November 2011 Q6

12 The equation of a curve is $\dfrac{e^{2x}}{1+e^{2x}}$. Show that the gradient of the curve at the point for which $x = \ln 3$ is $\dfrac{9}{50}$. [4]

Cambridge International AS & A Level Mathematics, 9709/33 November 2011 Q2

5 Integration

1 (i) Given that $y = \tan 2x$, find $\dfrac{dy}{dx}$. [2]

(ii) Hence, or otherwise, show that
$$\int_0^{\frac{1}{6}\pi} \sec^2 2x \, dx = \tfrac{1}{2}\sqrt{3},$$
and, by using an appropriate trigonometrical identity, find the exact value of
$$\int_0^{\frac{1}{6}\pi} \tan^2 2x \, dx.$$ [6]

(iii) Use the identity $\cos 4x \equiv 2\cos^2 2x - 1$ to find the exact value of $\displaystyle\int_0^{\frac{1}{6}\pi} \dfrac{1}{1+\cos 4x} \, dx$. [2]

Cambridge International AS & A Level Mathematics, 9709/02 November 2006 Q7

2 (i) (a) Prove the identity
$$\sec^2 x + \sec x \tan x \equiv \frac{1+\sin x}{\cos^2 x}.$$

(b) Hence prove that
$$\sec^2 x + \sec x \tan x \equiv \frac{1}{1-\sin x}.$$ [3]

(ii) By differentiating $\dfrac{1}{\cos x}$, show that if $y = \sec x$ then $\dfrac{dy}{dx} = \sec x \tan x$. [3]

(iii) Using the results of parts (i) and (ii), find the exact value of

$$\int_0^{\frac{1}{4}\pi} \dfrac{1}{1-\sin x}\,dx.$$ [3]

Cambridge International AS & A Level Mathematics, 9709/02 November 2008 Q8

3 The diagram shows the curve $y = \sqrt{(1 + 2\tan^2 x)}$ for $0 \leq x \leq \frac{1}{4}\pi$.

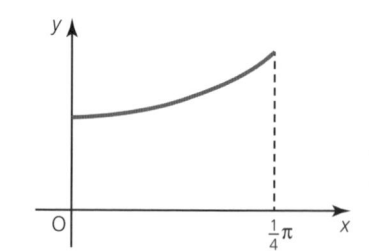

(i) Use the trapezium rule with three intervals to estimate the value of

$$\int_0^{\frac{1}{4}\pi} \sqrt{(1 + 2\tan^2 x)}\,dx,$$

giving your answer correct to 2 decimal places. [3]

(ii) The estimate found in part (i) is denoted by E. Explain, without further calculation, whether another estimate found using the trapezium rule with six intervals would be greater than E or less than E. [1]

Cambridge International AS & A Level Mathematics, 9709/3 May/June 2009 Q2

4 (i) Prove the identity $\cos 4\theta - 4\cos 2\theta + 3 \equiv 8\sin^4\theta$. [4]

(ii) Using this result find, in simplified form, the exact value of

$$\int_{\frac{1}{6}\pi}^{\frac{1}{3}\pi} \sin^4\theta \, d\theta.$$ [4]

Cambridge International AS & A Level Mathematics, 9709/31 November 2009 Q5

5 (i) By differentiating $\dfrac{\cos x}{\sin x}$, show that if $y = \cot x$ then $\dfrac{dy}{dx} = -\text{cosec}^2 x$. [3]

(ii) By expressing $\cot^2 x$ in terms of $\text{cosec}^2 x$ and using the result of part **(i)**, show that $\displaystyle\int_{\frac{1}{4}\pi}^{\frac{1}{2}\pi} \cot^4 x \, dx = 1 - \tfrac{1}{4}\pi$. [4]

(iii) Express $\cos 2x$ in terms of $\sin^2 x$ and hence show that $\dfrac{1}{1-\cos 2x}$ can be expressed as $\tfrac{1}{2}\operatorname{cosec}^2 x$.

Hence, using the result of part (i), find

$$\int \dfrac{1}{1-\cos 2x}\,dx.$$ [3]

Cambridge International AS & A Level Mathematics, 9709/21 May/June 2010 Q8

6 (i) Using the expansions of $\cos(3x - x)$ and $\cos(3x + x)$, prove that

$$\tfrac{1}{2}(\cos 2x - \cos 4x) \equiv \sin 3x \sin x.$$ [3]

(ii) Hence show that $\displaystyle\int_{\frac{1}{6}\pi}^{\frac{1}{3}\pi} \sin 3x \sin x \, dx = \tfrac{1}{8}\sqrt{3}$. [3]

Cambridge International AS & A Level Mathematics, 9709/31 May/June 2010 Q4

7 The diagram shows the curve $y = \sqrt{(1 + x^3)}$.

Region A is bounded by the curve and the lines $x = 0$, $x = 2$ and $y = 0$.

Region B is bounded by the curve and the lines $x = 0$ and $y = 3$.

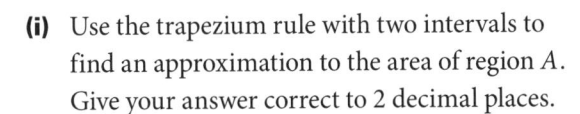

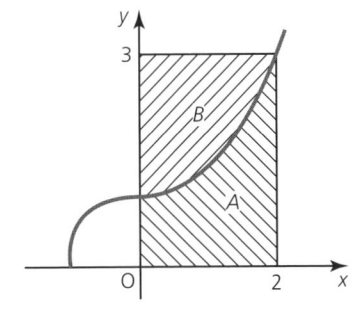

(i) Use the trapezium rule with two intervals to find an approximation to the area of region A. Give your answer correct to 2 decimal places. [3]

(ii) Deduce an approximation to the area of region B and explain why this approximation under-estimates the true area of region B. [2]

Cambridge International AS & A Level Mathematics, 9709/22 May/June 2011 Q2

8 Find the exact value of the positive constant k for which

$$\int_0^k e^{4x}\,dx = \int_0^{2k} e^x\,dx.$$ [6]

Cambridge International AS & A Level Mathematics, 9709/22 November 2011 Q4

6 Numerical solution of equations

1 (i) By sketching a suitable pair of graphs, show that the equation
$$2 - x = \ln x$$
has only one root. [2]

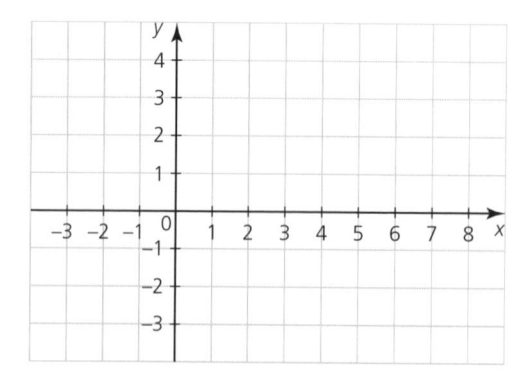

(ii) Verify by calculation that this root lies between 1.4 and 1.7. [2]

(iii) Show that this root also satisfies the equation
$$x = \tfrac{1}{3}(4 + x - 2\ln x).$$ [1]

(iv) Use the iterative formula
$$x_{n+1} = \tfrac{1}{3}(4 + x_n - 2\ln x_n),$$
with initial value $x_1 = 1.5$, to determine this root correct to 2 decimal places. Give the result of each iteration to 4 decimal places. [3]

Cambridge International AS & A Level Mathematics, 9709/03 November 2007 Q6

2 The diagram shows the curve $y = x^2 \cos x$, for $0 \leq x \leq \frac{1}{2}\pi$, and its maximum point M.

   (i) Show by differentiation that the x-coordinate of M satisfies the equation $\tan x = \dfrac{2}{x}$. [4]

   (ii) Verify by calculation that this equation has a root (in radians) between 1 and 1.2. [2]

   (iii) Use the iterative formula $x_{n+1} = \tan^{-1}\left(\dfrac{2}{x_n}\right)$ to determine this root correct to 2 decimal places. Give the result of each iteration to 4 decimal places. [3]

   Cambridge International AS & A Level Mathematics, 9709/22 November 2009 Q7

3 (i) By sketching suitable graphs, show that the equation

$$4x^2 - 1 = \cot x$$

has only one root in the interval $0 < x < \frac{1}{2}\pi$. [2]

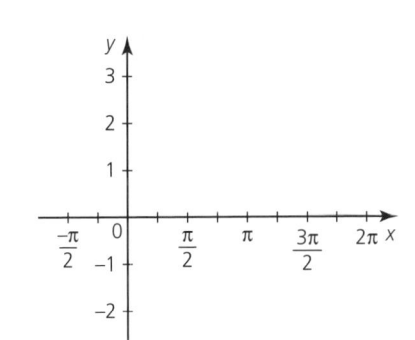

(ii) Verify by calculation that this root lies between 0.6 and 1. [2]

(iii) Use the iterative formula

$$x_{n+1} = \tfrac{1}{2}\sqrt{(1 + \cot x_n)}$$

to determine the root correct to 2 decimal places. Give the result of each iteration to 4 decimal places. [3]

Cambridge International AS & A Level Mathematics, 9709/31 November 2010 Q4

4 The sequence of values given by the iterative formula

$$x_{n+1} = \frac{7x_n}{8} + \frac{5}{2x_n^4},$$

with initial value $x_1 = 1.7$, converges to α.

(i) Use this iterative formula to determine α correct to 2 decimal places, giving the result of each iteration to 4 decimal places. [3]

(ii) State an equation that is satisfied by α and hence show that $\alpha = \sqrt[5]{20}$. [2]

Cambridge International AS & A Level Mathematics, 9709/23 November 2010 Q2

5 The diagram shows a semicircle ACB with centre O and radius r. The tangent at C meets AB produced at T. The angle BOC is x radians. The area of the shaded region is equal to the area of the semicircle.

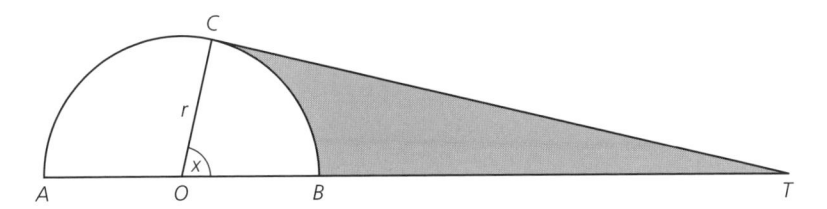

(i) Show that x satisfies the equation

$$\tan x = x + \pi.$$ [3]

(ii) Use the iterative formula $x_{n+1} = \tan^{-1}(x_n + \pi)$ to determine x correct to 2 decimal places. Give the result of each iteration to 4 decimal places. [3]

Cambridge International AS & A Level Mathematics, 9709/32 May/June 2011 Q4

6 (i) By sketching a suitable pair of graphs, show that the equation

$$\frac{1}{x} = \sin x,$$

where x is in radians, has only one root for

$0 < x \leqslant \tfrac{1}{2}\pi.$ [2]

(ii) Verify by calculation that this root lies between $x = 1.1$ and $x = 1.2$. [2]

(iii) Use the iterative formula $x_{n+1} = \dfrac{1}{\sin x_n}$ to determine this root correct to 2 decimal places. Give the result of each iteration to 4 decimal places. [3]

Cambridge International AS & A Level Mathematics, 9709/22 November 2011 Q5

7 Further algebra

1 Let $f(x) = \dfrac{x^3 - x - 2}{(x-1)(x^2+1)}$.

  (i) Express $f(x)$ in the form
  $$A + \dfrac{B}{x-1} + \dfrac{Cx+D}{x^2+1},$$
  where A, B, C and D are constants. [5]

  (ii) Hence show that $\int_2^3 f(x)\,dx = 1$. [4]

Cambridge International AS & A Level Mathematics, 9709/03 November 2003 Q8

2 Let $f(x) = \dfrac{x^2 + 7x - 6}{(x-1)(x-2)(x+1)}$.

  (i) Express $f(x)$ in partial fractions. [4]

(ii) Show that, when x is sufficiently small for x^4 and higher powers to be neglected,

$f(x) = -3 + 2x - \frac{3}{2}x^2 + \frac{11}{4}x^3$. [5]

Cambridge International AS & A Level Mathematics, 9709/03 May/June 2004 Q9

3 An appropriate form for expressing $\dfrac{3x}{(x+1)(x-2)}$ in partial fractions is

$\dfrac{A}{x+1} + \dfrac{B}{x-2}$,

where A and B are constants.

(a) Without evaluating any constants, state appropriate forms for expressing the following in partial fractions:

(i) $\dfrac{4x}{(x+4)(x^2+3)}$, [1]

(ii) $\dfrac{2x+1}{(x-2)(x+2)^2}$ [2]

(b) Show that $\int_3^4 \dfrac{3x}{(x+1)(x-2)}\,dx = \ln 5$. [6]

Cambridge International AS & A Level Mathematics, 9709/03 November 2004 Q8

4 Expand $\dfrac{1}{(2+x)^3}$ in ascending powers of x, up to and including the term in x^2, simplifying the coefficients. [4]

Cambridge International AS & A Level Mathematics, 9709/03 November 2004 Q1

5 Expand $(1+4x)^{-\frac{1}{2}}$ in ascending powers of x, up to and including the term in x^3, simplifying the coefficients. [4]

Cambridge International AS & A Level Mathematics, 9709/03 May/June 2005 Q1

6 (i) Express $\dfrac{5x-x^2}{(1+x)(2+x^2)}$ in partial fractions. [5]

(ii) Hence obtain the expansion of $\dfrac{5x-x^2}{(1+x)(2+x^2)}$ in ascending powers of x, up to and including the term in x^3. [5]

Cambridge International AS & A Level Mathematics, 9709/32 May/June 2011 Q8

7 The polynomial p(x) is defined by

$$p(x) = ax^3 - x^2 + 4x - a,$$

where a is a constant. It is given that $(2x - 1)$ is a factor of p(x).

(i) Find the value of a and hence factorise p(x). [4]

(ii) When *a* has the value found in part **(i)**, express $\dfrac{8x-13}{\text{p}(x)}$ in partial fractions. [5]

Cambridge International AS & A Level Mathematics, 9709/33 November 2011 Q7

8 Further integration

1 The diagram shows the curve $y = x^2\sqrt{(1-x^2)}$ for $x \geqslant 0$ and its maximum point M.

  (i) Show, by means of the substitution $x = \sin\theta$, that the area A of the shaded region between the curve and the x-axis is given by

  $$A = \tfrac{1}{4}\int_0^{\frac{1}{2}\pi} \sin^2 2\theta \, d\theta.$$ [3]

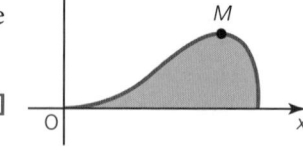

  (ii) Hence obtain the exact value of A. [4]

Cambridge International AS & A Level Mathematics, 9709/3 May/June 2009 Q10 (ii),(iii)

2 (i) Express $\dfrac{2}{(x+1)(x+3)}$ in partial fractions. [2]

  (ii) Using your answer to part **(i)**, show that

  $$\left(\dfrac{2}{(x+1)(x+3)}\right)^2 \equiv \dfrac{1}{(x+1)^2} - \dfrac{1}{x+1} + \dfrac{1}{x+3} + \dfrac{1}{(x+3)^2}.$$ [2]

(iii) Hence show that $\int_0^1 \dfrac{4}{(x+1)^2(x+3)^2}\,dx = \dfrac{7}{12} - \ln\dfrac{3}{2}$. [5]

Cambridge International AS & A Level Mathematics, 9709/31 May/June 2010 Q8

3 The diagram shows the curve $y = x^3 \ln x$ and its minimum point M.

Find the exact area of the shaded region bounded by the curve, the x-axis and the line $x = 2$. [5]

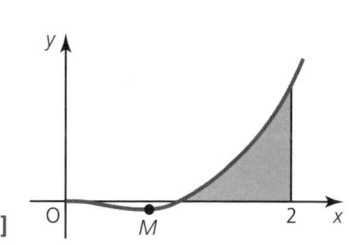

Cambridge International AS & A Level Mathematics, 9709/31 November 2010 Q9(ii)

4 Let $I = \int_0^1 \dfrac{x^2}{\sqrt{(4-x^2)}}\,dx$.

(i) Using the substitution $x = 2\sin\theta$, show that
$$I = \int_0^{\frac{1}{6}\pi} 4\sin^2\theta\,d\theta.$$
[3]

(ii) Hence find the exact value of I. [4]

....

....

....

....

....

Cambridge International AS & A Level Mathematics, 9709/31 November 2010 Q5

5 The diagram shows the curve $y = x^2 e^{-x}$.

   (i) Show that the area of the shaded region bounded by the curve, the x-axis and the line $x = 3$ is equal to $2 - \dfrac{17}{e^3}$. [5]

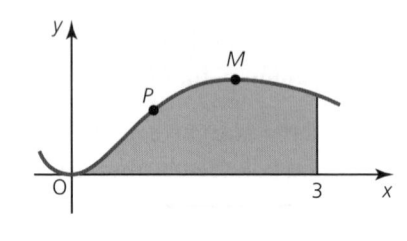

....

....

....

....

   (ii) Find the x-coordinate of the maximum point M on the curve. [4]

....

....

   (iii) Find the x-coordinate of the point P at which the tangent to the curve passes through the origin. [2]

....

....

....

....

Cambridge International AS & A Level Mathematics, 9709/32 May/June 2011 Q10

6 (i) Use the substitution $u = \tan x$ to show that, for $n \neq -1$,

$$\int_0^{\frac{1}{4}\pi} (\tan^{n+2} x + \tan^n x)\,dx = \frac{1}{n+1}.$$ [4]

(ii) Hence find the exact value of

(a) $\int_0^{\frac{1}{4}\pi} (\sec^4 x - \sec^2 x)\,dx$, [3]

(b) $\int_0^{\frac{1}{4}\pi} (\tan^9 x + 5\tan^7 x + 5\tan^5 x + \tan^3 x)\,dx$. [3]

Cambridge International AS & A Level Mathematics, 9709/33 November 2011 Q10

9 Differential equations

1 A rectangular reservoir has a horizontal base of area $1000 \, \text{m}^2$. At time $t = 0$, it is empty and water begins to flow into it at a constant rate of $30 \, \text{m}^3 \, \text{s}^{-1}$. At the same time, water begins to flow out at a rate proportional to $\sqrt{h}$, where h m is the depth of the water at time t s. When $h = 1$, $\dfrac{dh}{dt} = 0.02$.

(i) Show that h satisfies the differential equation

$$\frac{dh}{dt} = 0.01(3 - \sqrt{h}).$$ [3]

It is given that, after making the substitution $x = 3 - \sqrt{h}$, the equation in part **(i)** becomes $(x - 3) \dfrac{dx}{dt} = 0.005x$.

(ii) Using the fact that $x = 3$ when $t = 0$, solve this differential equation, obtaining an expression for t in terms of x. [5]

(iii) Find the time at which the depth of water reaches $4 \, \text{m}$. [2]

Cambridge International AS & A Level Mathematics, 9709/03 November 2004 Q10

2 In a certain industrial process, a substance is being produced in a container. The mass of the substance in the container t minutes after the start of the process is x grams. At any time, the rate of formation of the substance is proportional to its mass. Also, throughout the process, the substance is removed from the container at a constant rate of 25 grams per minute. When $t = 0$, $x = 1000$ and $\frac{dx}{dt} = 75$.

(i) Show that x and t satisfy the differential equation
$$\frac{dx}{dt} = 0.1(x - 250).$$ [2]

(ii) Solve this differential equation, obtaining an expression for x in terms of t. [6]

Cambridge International AS & A Level Mathematics, 9709/03 May/June 2006 Q5

3 A model for the height, h metres, of a certain type of tree at time t years after being planted assumes that, while the tree is growing, the rate of increase in height is proportional to $(9-h)^{\frac{1}{3}}$. It is given that, when $t = 0$, $h = 1$ and $\frac{dh}{dt} = 0.2$.

(i) Show that h and t satisfy the differential equation
$$\frac{dh}{dt} = 0.1\,(9-h)^{\frac{1}{3}}.$$ [2]

(ii) Solve this differential equation, and obtain an expression for h in terms of t. [7]

(iii) Find the maximum height of the tree and the time taken to reach this height after planting. [2]

(iv) Calculate the time taken to reach half the maximum height. [1]

Cambridge International AS & A Level Mathematics, 9709/03 May/June 2007 Q10

4 The number of insects in a population t days after the start of observations is denoted by N. The variation in the number of insects is modelled by a differential equation of the form

$$\frac{dN}{dt} = kN\cos(0.02t),$$

where k is a constant and N is taken to be a continuous variable. It is given that $N = 125$ when $t = 0$.

(i) Solve the differential equation, obtaining a relation between N, k and t. [5]

(ii) Given also that $N = 166$ when $t = 30$, find the value of k. [2]

(iii) Obtain an expression for N in terms of t, and find the least value of N predicted by this model. [3]

Cambridge International AS & A Level Mathematics, 9709/03 November 2007 Q7

5 In the diagram the tangent to a curve at a general point P with co-ordinates (x, y) meets the x-axis at T. The point N on the x-axis is such that PN is perpendicular to the x-axis.

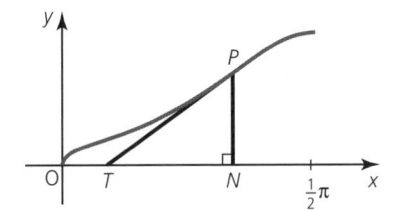

The curve is such that, for all values of x in the interval $0 < x < \tfrac{1}{2}\pi$, the area of triangle PTN is equal to $\tan x$, where x is in radians.

(i) Using the fact that the gradient of the curve at P is $\dfrac{PN}{TN}$, show that

$$\dfrac{dy}{dx} = \tfrac{1}{2} y^2 \cot x.$$ [3]

(ii) Given that $y = 2$ when $x = \tfrac{1}{6}\pi$, solve this differential equation to find the equation of the curve, expressing y in terms of x. [6]

Cambridge International AS & A Level Mathematics, 9709/03 May/June 2008 Q8

6 In a model of the expansion of a sphere of radius r cm, it is assumed that, at time t seconds after the start, the rate of increase of the surface area of the sphere is proportional to its volume. When $t = 0$, $r = 5$ and $\dfrac{dr}{dt} = 2$.

   (i) Show that r satisfies the differential equation
   $$\dfrac{dr}{dt} = 0.08r^2.$$
   [4]

   [The surface area A and volume V of a sphere of radius r are given by the formulae $A = 4\pi r^2$, $V = \tfrac{4}{3}\pi r^3$.]

   (ii) Solve this differential equation, obtaining an expression for r in terms of t. [5]

   (iii) Deduce from your answer to part (ii) the set of values that t can take, according to this model. [1]

   Cambridge International AS & A Level Mathematics, 9709/31 November 2009 Q10

7 A certain curve is such that its gradient at a point (x, y) is proportional to xy. At the point $(1, 2)$ the gradient is 4.

   (i) By setting up and solving a differential equation, show that the equation of the curve is $y = 2e^{x^2 - 1}$. [7]

(ii) State the gradient of the curve at the point (−1, 2) and sketch the curve. [2]

Cambridge International AS & A Level Mathematics, 9709/32 May/June 2011 Q6

8 During an experiment, the number of organisms present at time t days is denoted by N, where N is treated as a continuous variable. It is given that

$$\frac{dN}{dt} = 1.2e^{-0.02t}N^{0.5}.$$

When $t = 0$, the number of organisms present is 100.

(i) Find an expression for N in terms of t. [6]

(ii) State what happens to the number of organisms present after a long time. [1]

Cambridge International AS & A Level Mathematics, 9709/33 November 2011 Q4

P3

10 Vectors

1 The lines l and m have vector equations

$\mathbf{r} = \mathbf{i} - 2\mathbf{k} + s(2\mathbf{i} + \mathbf{j} + 3\mathbf{k})$ and $\mathbf{r} = 6\mathbf{i} - 5\mathbf{j} + 4\mathbf{k} + t(\mathbf{i} - 2\mathbf{j} + \mathbf{k})$

respectively.

(i) Show that l and m intersect, and find the position vector of their point of intersection. [5]

(ii) Find the equation of the plane containing l and m, giving your answer in the form $ax + by + cz = d$. [6]

Cambridge International AS & A Level Mathematics, 9709/03 November 2003 Q10

2 With respect to the origin O, the points P, Q, R, S have position vectors given by

$\overrightarrow{OP} = \mathbf{i} - \mathbf{k}, \quad \overrightarrow{OQ} = -2\mathbf{i} + 4\mathbf{j}, \quad \overrightarrow{OR} = 4\mathbf{i} + 2\mathbf{j} + \mathbf{k}, \quad \overrightarrow{OS} = 3\mathbf{i} + 5\mathbf{j} - 6\mathbf{k}.$

(i) Find the equation of the plane containing P, Q and R, giving your answer in the form $ax + by + cz = d$. [6]

(ii) The point N is the foot of the perpendicular from S to this plane. Find the position vector of N and show that the length of SN is 7. [6]

Cambridge International AS & A Level Mathematics, 9709/03 May/June 2004 Q11

3 The lines l and m have vector equations
$\mathbf{r} = 2\mathbf{i} - \mathbf{j} + 4\mathbf{k} + s(\mathbf{i} + \mathbf{j} - \mathbf{k})$ and $\mathbf{r} = -2\mathbf{i} + 2\mathbf{j} + \mathbf{k} + t(-2\mathbf{i} + \mathbf{j} + \mathbf{k})$
respectively.

(i) Show that l and m do not intersect. [4]

The point P lies on l and the point Q has position vector $2\mathbf{i} - \mathbf{k}$.

(ii) Given that the line PQ is perpendicular to l, find the position vector of P. [4]

(iii) Verify that Q lies on m and that PQ is perpendicular to m. [2]

Cambridge International AS & A Level Mathematics, 9709/03 November 2004 Q9

4 The straight line l passes through the points A and B with position vectors

$2\mathbf{i} + 2\mathbf{j} + \mathbf{k}$ and $\mathbf{i} + 4\mathbf{j} + 2\mathbf{k}$

respectively. This line intersects the plane p with equation $x - 2y + 2z = 6$ at the point C.

(i) Find the position vector of C. [4]

(ii) Find the acute angle between l and p. [4]

(iii) Show that the perpendicular distance from A to p is equal to 2. [3]

Cambridge International AS & A Level Mathematics, 9709/03 November 2005 Q10

5 With respect to the origin O, the points A, B and C have position vectors given by

$\overrightarrow{OA} = \mathbf{i} - \mathbf{k}$, $\overrightarrow{OB} = 3\mathbf{i} + 2\mathbf{j} - 3\mathbf{k}$ and $\overrightarrow{OC} = 4\mathbf{i} - 3\mathbf{j} + 2\mathbf{k}$.

The mid-point of AB is M. The point N lies on AC between A and C and is such that $AN = 2NC$.

(i) Find a vector equation of the line MN. [4]

(ii) It is given that MN intersects BC at the point P. Find the position vector of P. [4]

Cambridge International AS & A Level Mathematics, 9709/31 November 2009 Q6

6 The lines l and m have vector equations

$\mathbf{r} = \mathbf{i} + \mathbf{j} + \mathbf{k} + s(\mathbf{i} - \mathbf{j} + 2\mathbf{k})$ and $\mathbf{r} = 4\mathbf{i} + 6\mathbf{j} + \mathbf{k} + t(2\mathbf{i} + 2\mathbf{j} + \mathbf{k})$

respectively.

(i) Show that l and m intersect. [4]

(ii) Calculate the acute angle between the lines. [3]

(iii) Find the equation of the plane containing l and m, giving your answer in the form $ax + by + cz = d$. [5]

Cambridge International AS & A Level Mathematics, 9709/31 May/June 2010 Q10

7 With respect to the origin O, the points A and B have position vectors given by $\overrightarrow{OA} = \mathbf{i} + 2\mathbf{j} + 2\mathbf{k}$ and $\overrightarrow{OB} = 3\mathbf{i} + 4\mathbf{j}$. The point P lies on the line AB and OP is perpendicular to AB.

(i) Find a vector equation for the line AB. [1]

(ii) Find the position vector of P. [4]

(iii) Find the equation of the plane which contains AB and which is perpendicular to the plane OAB, giving your answer in the form $ax + by + cz = d$. [4]

Cambridge International AS & A Level Mathematics, 9709/31 November 2010 Q7

8 The line l has equation $\mathbf{r} = \begin{pmatrix} a \\ 1 \\ 4 \end{pmatrix} + \lambda \begin{pmatrix} 4 \\ 3 \\ -2 \end{pmatrix}$, where a is a constant.

The plane p has equation $2x - 2y + z = 10$.

(i) Given that l does not lie in p, show that l is parallel to p. [2]

(ii) Find the value of a for which l lies in p. [2]

(iii) It is now given that the distance between l and p is 6. Find the possible values of a. [5]

Cambridge International AS & A Level Mathematics, 9709/33 November 2011 Q9

11 Complex numbers

1 The complex number u is given by $u = \dfrac{7+4i}{3-2i}$.

   (i) Express u in the form $x + iy$, where x and y are real. [3]

   (ii) Sketch an Argand diagram showing the point representing the complex number u. Show on the same diagram the locus of the complex number z such that $|z - u| = 2$. [3]

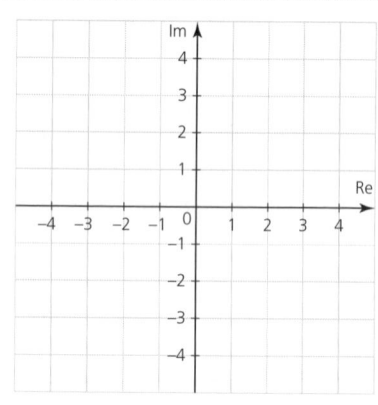

   (iii) Find the greatest value of arg z for points on this locus. [3]

Cambridge International AS & A Level Mathematics, 9709/03 November 2003 Q7

2 (i) Find the roots of the equation $z^2 - z + 1 = 0$, giving your answers in the form $x + iy$, where x and y are real. [2]

   (ii) Obtain the modulus and argument of each root. [3]

(iii) Show that each root also satisfies the equation $z^3 = -1$ [2]

....

....

....

Cambridge International AS & A Level Mathematics, 9709/03 May/June 2004 Q8

3 The complex numbers $1 + 3i$ and $4 + 2i$ are denoted by u and v respectively.

   (i) Find, in the form $x + iy$, where x and y are real, the complex numbers $u - v$ and $\dfrac{u}{v}$. [3]

....

....

....

   (ii) State the argument of $\dfrac{u}{v}$. [1]

....

In an Argand diagram, with origin O, the points A, B and C represent the numbers u, v and $u - v$ respectively.

   (iii) State fully the geometrical relationship between OC and BA. [2]

....

   (iv) Prove that angle $AOB = \frac{1}{4}\pi$ radians. [2]

....

Cambridge International AS & A Level Mathematics, 9709/03 November 2004 Q6

4 The complex number $-2 + i$ is denoted by u.

   (i) Given that u is a root of the equation $x^3 - 11x - k = 0$, where k is real, find the value of k. [3]

....

....

....

....

(ii) Write down the other complex root of this equation. [1]

(iii) Find the modulus and argument of u. [2]

(iv) Sketch an Argand diagram showing the point representing u. Shade the region whose points represent the complex numbers z satisfying both the inequalities

$|z| < |z - 2|$ and $0 < \arg(z - u) < \frac{1}{4}\pi$. [4]

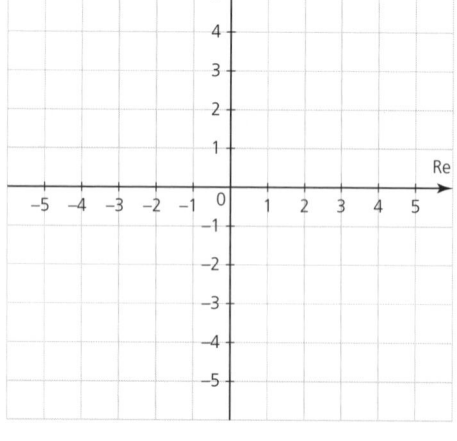

Cambridge International AS & A Level Mathematics, 9709/31 November 2009 Q7

5 The complex number $2 + 2i$ is denoted by u.

(i) Find the modulus and argument of u. [2]

(ii) Sketch an Argand diagram showing the points representing the complex numbers 1, i and u. Shade the region whose points represent the complex numbers z which satisfy both the inequalities $|z - 1| \leq |z - i|$ and $|z - u| \leq 1$. [4]

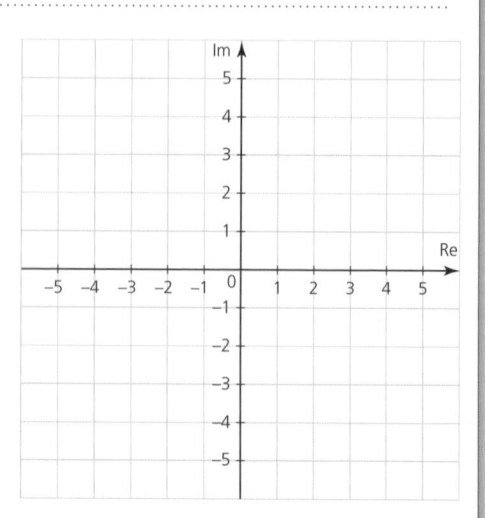

(iii) Using your diagram, calculate the value of $|z|$ for the point in this region for which arg z is least. [3]

Cambridge International AS & A Level Mathematics, 9709/31 May/June 2010 Q7

6 The complex number w is defined by $w = -1 + i$.

(i) Find the modulus and argument of w^2 and w^3, showing your working. [4]

(ii) The points in an Argand diagram representing w and w^2 are the ends of a diameter of a circle. Find the equation of the circle, giving your answer in the form $|z - (a + bi)| = k$. [4]

Cambridge International AS & A Level Mathematics, 9709/33 November 2011 Q6